SEEKING 1

Seeking the Greatest Good

THE CONSERVATION LEGACY OF GIFFORD PINCHOT

Char Miller

*To Jennifer —
with so many
Thanks for
you work at
PIC !*

Char

UNIVERSITY OF PITTSBURGH PRESS

9-20-2013

Grey Towers

Published by the University of Pittsburgh Press, Pittsburgh, Pa., 15260

Copyright © 2013, University of Pittsburgh Press

Manufactured in the United States of America

Printed on acid-free paper

10 9 8 7 6 5 4 3 2 1

Library of Congress Cataloging-in-Publication Data

Miller, Char, 1951–
Seeking the Greatest Good: The Conservation Legacy of Gifford Pinchot
/ Char Miller.
    pages cm
    Includes bibliographical references and index.
    ISBN 978-0-8229-6267-0 (paperback: acid-free paper)
    1. Pinchot Institute for Conservation—History. 2. Pinchot Institute for
Conservation—Influence. 3. Environmental policy—United States—History.
4. Nature conservation—United States—History. 5. Forest conservation—
United States—History. 6. Environmental education—United States—
History. 7. Pinchot, Gifford, 1865–1946—Influence. 8. Pinchot, Gifford,
1865–1946—Family. I. Title.
    QH76.M535 2013
    333.72—dc23   2013023785

CONTENTS

ACKNOWLEDGMENTS

Sitting at a well-used desk in one of the offices located on the third floor of Grey Towers National Historical Site and overhearing visitors as they strolled around the flower-filled grounds, commented on the lovingly restored bluestone mansion, or pointed out landmarks set within the site's commanding views east across the Delaware River valley, I knew just how privileged my position was. I was there to write a book about the estate, the familial home of the conservationist Gifford Pinchot, and the institute that bears his name, and as I worked my way through the books and documents filed behind glass-fronted cabinets, rifled through other materials scattered in this collection and in the many boxes that constitute the site's formal archive, even then I realized how deeply indebted I was (and would become) to a host of colleagues and friends who made this project possible.

At Grey Towers itself I have had an extraordinary opportunity to work with a series of its directors, many of whom make an appearance in this text, and have learned from and leaned on them since my first visit to the site in 1989. Most crucial to this project has been Edgar Brannon, a former director who shared his memories and personal diaries of his years at the site that have indispensably deepened my understanding of its import; the generosity of Allison Stewart, the current director, did much to smooth my research and writing. The archivist Becky Philpot was indefatigable in responding to my many questions and unearthing relevant objects, photographs, and correspondence with astonishing speed, including the creation of a digital archive. Lori McKean, Molly Breitbach, and the rest of the staff were beyond kind in letting me interrupt their days with my queries. The same is true of former staffers Gary Hines, Kimo Kimokeo, Carol Severance, and Amy Snyder, whose contributions to and remembrances of Grey

Towers added immeasurably to my understanding of its complex history and their part in its making.

Al Sample, the president of the Pinchot Institute for Conservation, had the idea for this book to explore the twined organizations' history and mark their fiftieth anniversaries. He too was patient with the endless stream of emails and phone interviews and was a fount of insight and information and was a most careful reader of the manuscript. A small grant from the Pinchot Institute for Conservation—the Edgar Brannon Conservation Fellowship—enabled me to spend ten writing-filled days at Grey Towers in the summer of 2012, for which I am grateful and doubly so for the aid that the institute's staff, including Jennifer Yeager, William Price, Alex Andrus, Stephanie Dalke, and Shannon Sutherland, gave me every step of the way.

For anyone who studies the U.S. Forest Service and the larger profession of forestry here and abroad, the Forest History Society, located in Durham, North Carolina, is the essential archive. With funding from an Alfred T. Bell Travel Grant, I was able to spend an intense week roaming through its collections with the brilliant librarian Cheryl Oakes. Rendering invaluable support too was Steven Anderson, president and CEO of the organization, James G. Lewis, Eben Lehman, and Andrea Anderson.

Some of the most stimulating conversations I have had about this and related projects, and the wider context in which the American environmental movement operates, have occurred with various members of the Pinchot family. Gifford Pinchot III, Peter Pinchot, Nancy Pittman Pinchot, Leila Pinchot, and Bibi Gaston have been engaged colleagues and constructive critics.

Supportive too were Michael Knies and Elizabeth Teets at the Weinberg Memorial Library of the University of Scranton, home to Rep. Joseph McDade's papers, as well as Philip Roberts, Tamsen Hert, and Andrea Binder, who directed me to relevant documents and correspondence in the Samuel Ordway Papers at the American History Center at the University of Wyoming. Closer to home, I am grateful to the Pomona College Research Committee for its support. I have also continuously relied on the skill and resourcefulness of Carrie Marsh and Lisa Crane in Special Collections at the Honnold/Mudd Library of the Claremont Colleges and their stellar colleagues (and mine) Char Booth and Sean Stone. The baristas at the Honnold Café, where much of this book was drafted, were generous with their pours.

At the University of Pittsburgh Press I have benefited from the wise counsel, smart suggestions, and good friendship of Cynthia Miller, now former director of the press; her retirement in early 2013 is a great loss for scholars everywhere! The press's staff, as in the past, has been a model of professionalism as were the anonymous reviewers' constructive criticisms. They have all made this a much better book.

Making the process a lot of fun too has been my family. Sandwiched in between my research trips to Grey Towers and the Forest History Society was a memorable three-generational gathering in Cape May, New Jersey, in July 2012. Disrupted at its start by the epic derecho that ripped up trees, shut down transportation, and darkened much of the mid-Atlantic, Judi and I nonetheless reveled in the time spent hanging out with our children and grandchildren, uplifted by their energy. A special shout out to our son Ben who has been a source of great advice on all things congressional and who works on the Hill, a political environment that has been, and remains, critical to the operations of Grey Towers and the Pinchot Institute—and to anyone who wants to understand them.

SEEKING THE GREATEST GOOD

Introduction

On a warm Friday morning in late June 2012, a party of volunteers—mostly board members of the Pinchot Institute for Conservation and staffers from Grey Towers National Historic Site—put blade to ground on the Jorritsma family's century-old dairy farm in Sussex County, New Jersey. Within minutes, they had dug a series of deep, round holes along the western bank of the Paulins Kill. As they planted willows and silky dogwood in the floodplain, still spongy underfoot from a recent storm, they could begin to see what the program's advocates had in mind years earlier when they conceived the Paulins Kill restoration project—a well-wooded terrain through which a sparkling river meanders on its way to the Delaware River and the ocean beyond.

Building the collaborative energy to launch this project proved less fluid. In 1994 the Sussex County Municipal Utilities Authority had taken the lead in developing a countywide plan in concert with local stakeholders to restore, maintain, and enhance local waterways, a process that received state funding six years later. Out of this initial activity emerged the Wallkill River Watershed Management Group, which was charged with coordinating and facilitating the restoration activities within the Upper Paulins Kill watershed. In 2008 the group's planners, ecologists, and landscape specialists met with the Jorritsma family, third-generation farmers whose herd of two thousand purebred Guernsey cattle had heavily grazed along nearly a mile-long stretch of the Paulins Kill. Yet neither the Jorritsmas nor the

Sussex County Municipal Utilities Authority had the requisite funding to regenerate the forest buffer and stabilize the river's badly eroded banks.

That is when they turned to the Pinchot Institute for Conservation, which coordinates the Common Waters Partnership and Common Waters Fund, a nongovernmental initiative devoted to the restoration of the Upper Delaware River watershed. Maintaining this river's water quality is critical to the water supplies of New York City and Philadelphia, which is why Edgar Brannon, the former director of Grey Towers and one of the founders of Common Waters, has argued that the river "may be the most important freshwater resource east of the Mississippi." As represented in its tag line— "Clean Water, Healthy Forests, and Sustainable Communities"—Common Waters is a combination of ecological and human concerns, a forest-to-faucet approach to landscape restoration. Regenerating the Jorritsmas' riparian-buffer forest is one small step toward ensuring the high quality of water consumed by thirsty New Yorkers and Philadelphians.[1]

To underwrite this and similar projects, Common Waters and the Pinchot Institute have developed a network of private and corporate philanthropies, local, county, and state agencies, with support from the U.S. Forest Service and National Park Service. To promote restoration initiatives, these watershed projects have also received technical assistance and scientific data from the U.S. Fish and Wildlife Service, the Natural Resource Conservation Service, and the Nature Conservancy. Building such a complex array of partnerships is not easy, observed Nathaniel Sajdak, the director of the Wallkill River Watershed Management Group, but it is "one of the most important keys to successfully planning and implementing watershed restoration and protection strategies, initiatives, and projects."[2]

This story of collective action on behalf of people and the environments they inhabit is emblematic of the work that the Pinchot Institute for Conservation and its home base, Grey Towers National Historic Site, have been pursuing since their joint founding in 1963. These two institutions, and their twinned history of environmental engagement, are the central subject of *Seeking the Greatest Good*. Its first six chapters are framed around their collective creation, which was set in motion when Gifford Bryce Pinchot (1915–1989) donated his family's estate to the federal government, a gift that gained the blessing of the White House on September 24, 1963, when President John F. Kennedy came to Grey Towers to deliver the keynote address at the dedication of what was then called the Pinchot Institute for Conservation Studies at Grey Towers.

Yet neither this celebration nor the transfer of ownership would have occurred had not Grey Towers been the landscape most closely associated with the Forest Service's founding chief, Gifford Pinchot (1865–1946). Although the legendary forester and conservationist did not grow up at Grey Towers, his adult life revolved around the Norman chateau-like mansion that his parents, James and Mary Eno Pinchot, constructed in 1886. Indeed, the building was formally opened, and deliberately so, on the young man's twenty-first birthday. His family marked the august occasion by presenting Gifford with a gilt-edged copy of the bible of the fledgling conservation movement, George Perkins Marsh's *Man and Nature: Earth as Modified by Human Action.* Marsh argued that careful stewardship of natural resources was essential to the survival of modern industrial civilization, an argument that Pinchot hoped to exemplify through his chosen profession, forestry. He fulfilled that pledge as the first American-born scientifically trained forester in the United States, becoming the driving engine behind forestry's development as an academic discipline and professional practice. His work was particularly focused on the expansion and management of the national forest system, which now encompasses 193 million acres from coast to coast. When Pinchot was forced out of the Forest Service in 1910—President William Howard Taft fired him for insubordination—his career in public service did not end. From Grey Towers, Pinchot launched one political campaign after another, culminating in his two nonconsecutive terms as Pennsylvania's governor in the mid-1920s and early 1930s; during the latter term, he was credited with helping stabilize the state, then wracked by the Great Depression.[3]

Whether battling on the national or state level, Pinchot found in Grey Towers' sprawling grounds the perfect respite from the hurly-burly of daily life. A serious angler, he was never happier than when hip-deep in the Sawkill Creek that runs through the property, casting into its deep, dark pools. In important ways, his life was defined by his migrations between his professional occupations and political offices and his rural retreat in Milford.

With Pinchot's death in 1946 and, twelve years later, that of his wife, the human rights activist Cornelia Bryce Pinchot (1881–1960), the question of Grey Towers' future became a matter of pressing concern for the couple's only child, Gifford Bryce Pinchot. His decision to donate the house and surrounding grounds to the Forest Service set in motion the next stage of Grey Towers' evolution, from the Pinchot family's summer residence to a national landmark overseen by a federal agency. The final six chapters of *Seeking the Greatest Good* tracks this transformation over the succeeding half century through the intertwined histories of the mansion, the Pinchot Institute that was housed within it, the Forest Service charged with managing the historic structure, and three generations of the Pinchot family itself, who have remained deeply involved with the site.

This weave of institutional, administrative, and family history is also set within the context of U.S. environmental policy making since the mid-1960s. In his remarks at Grey Towers, President Kennedy described some of the immediate environmental challenges then facing the country: "The fact of the matter is that [the institute] is needed . . . more today than ever before, because we are reaching the limits of our fundamental need of water to drink, of fresh air to breathe, of open space to enjoy, of abundant sources of energy to make life easier." The president's insights helped define the Pinchot Institute's focus that ever since has been framed around such issues as clean air and clean water, endangered and threatened species, public lands management, habitat restoration, and environmental justice.

Born of an innovative partnership between the Forest Service and the Conservation Foundation (which later merged with the World Wildlife Fund), the Pinchot Institute has been also drawn to cooperative engagement as a form of organizational action. Yet the nature of its work has evolved, as have the goals, objectives, and strategies it has pursued. Dedicated one year after Rachel Carson's *Silent Spring* rocked the nation, the institute's initial aspiration was to use Grey Towers as a neutral site to encourage high-level conversations between leading scientists, educators, government officials, and activists over how to more ethically and efficiently manage America's natural resources. These linked purposes were essential, President Kennedy affirmed in his keynote address: "Today's conservation movement . . . must embrace disciplines scarcely known to its prophets of the past," and although government "must provide a national policy framework for this new conservation emphasis," to do so it requires "sound information, objective research and study. It is this function which the Pinchot Institute can serve most effectively."[4]

Yet the public-private partnership that launched the institute was not long sustained. By 1967 the Conservation Foundation had moved its headquarters from New York to Washington, D.C., and its priorities changed along with its address, leading to the collapse of its collaboration project with the Forest Service. In the aftermath the federal agency struggled to reconceive the institute, finally settling on a different form of alliance, this time with universities in the northeastern United States. In 1971, with Congress and the executive branch worried about the pressures confronting the dense urban corridor from Washington, D.C., to Boston, the Forest Service and its partners created the Pinchot Consortium to address some of these potentially explosive issues. Through this entity, an interdisciplinary cohort of researchers received small federal grants to analyze metropolitan air and water quality, conduct some of the first studies of acid rain fallout, and assess the ameliorative power of open space and vegetated landscapes on urbanites' health and happiness.

Despite the consortium's success, it fell victim to forces beyond its control. Beginning in the mid-1970s, presidents Carter and Reagan began to slash the federal budget—the latter indeed zeroed out funding for Grey Towers and the institute—and in 1984 the Forest Service took the hint, halting this productive cooperative arrangement. It would take nearly a decade for the Pinchot Institute to regain its footing. One major step in that process came when its leadership acknowledged the inherent vulnerability of a one-source funding model; no longer could the institute rely solely on the Forest Service's largesse. This realization led the institute's board of directors, which then included the chief of the Forest Service, to reorganize the institute as an independent nonprofit organization with the ability to seek additional financial support from charitable foundations and other private and public sources. The decision meant that the institute now had to demonstrate to funders and policy makers that its research and on-the-ground conservation projects are directly relevant to the information and action these entities have required. To reach this wider audience, the institute also decided it needed to establish a visible presence in the nation's capital in addition to its offices at Grey Towers.

While the history of Grey Towers and the Pinchot Institute reflect some of the formative issues driving the postwar environmental movement, Grey Towers as a physical setting illustrates another aspect of contemporary environmentalism. Beginning in the 1950s, as urban renewal projects bulldozed historic neighborhoods and suburbanization ran over rural communities, preservationists fought to protect and restore critical elements of the nation's built landscape. To secure sanction for their cause, they lobbied Congress for legislative means to defend imperiled structures and places. With the passage of the National Historic Preservation Act of 1966, activists gained an important tool with which to preserve properties deemed valuable examples of the national past, with the goal of fostering "conditions under which our modern society and our prehistoric and historic resources can exist in productive harmony." Sites of particular significance were designated National Historical Landmarks, and Grey Towers received that status in 1966 (it is now a National Historical Site). This newfound status did not guarantee Grey Towers a careful restoration. The Forest Service's initial rehabilitation projects tended to degrade the site's historic integrity, and it was not until 2001, when a nearly $20 million restoration project was completed, that the house and grounds recaptured some of their original grandeur.

The Pinchot Institute was at the same time undergoing a process of reinvention. Even as it continued its original mission to help the Forest Service make sense of emerging issues in natural resource management, it also responded to contemporary environmental debates that led it to craft policy responses to such contentious issues as wilderness protection, habitat regeneration, and the development of sustainable wood bioenergy. These efforts also have enmeshed the institute in the broader struggle to define the role of government in addressing environmental problems. One of them, the emergence of community forestry initiatives across the country and abroad, has led it to engage in the wider national and international dialogue about the need for inclusive, community-defined land stewardship practices. This contemporary activism has found historic legitimacy in its namesake's dictum: Management of the national forests, Gifford Pinchot asserted in 1901, "is primarily a local issue and should always be dealt with on local grounds. Local rules must be framed to meet local conditions, and they must be modified from time to time as local needs may require."[5]

Yet maintaining such a bottom-up perspective can be difficult for policy shops and think tanks headquartered in Washington, D.C. These independent research organizations got their start in the Progressive Era (and in 1909 Gifford Pinchot founded one of the first of them, the National Conservation Association), but they morphed in number and significance in the decades following the Second World War. The government's need "to marshal sophisticated technical expertise for both the Cold War national security enterprise and the short-lived war against poverty" led it to contract with an array of institutes, such as the Brookings Institution and RAND Corporation, for the requisite knowledge and professional guidance. During the late 1960s and early 1970s, a new breed of public interest groups emerged that were explicitly allied with conservative or liberal causes and the two major political parties and focused on providing analytical justification for specific policy agendas. As these entities raised money, lobbied Congress, and worked through the media to extend their influence, they helped build a so-called New Washington. In fact, they began to supplant political parties as a key ingredient in defining the capital's political life; their central function—the generation of a steady stream of independent, expert, and often partisan opinions—came at the same time that newspapers opened up their editorial pages for outside commentaries and network and cable television coverage expanded.

The Pinchot Institute has chosen to operate beneath the din. Its approach is a reflection of its historic commitment to nonpartisan analysis, a choice that is also a result of its founding mission, institutional history, and the professional aspirations of a staff attracted to its pragmatic approach to environmental policy making and practice. Consider the problem-solving strategy of one of its progenitors, the Conservation Foundation (1948). Through research and education, the foundation was committed to identifying how the nation could conserve resources and live more sustainably. However short-lived the relationship between it and the fledgling Pinchot Institute, the Conservation Foundation's articulation of its public role, and the social space it occupied—expertise offered to governments and citizens—has had an enduring influence on the institute's self-perception and self-representation.

Indeed, some of the Pinchot Institute's formative leadership came from the ranks of the Conservation Foundation. Its original governing board included the Conservation Foundation's founder Fairfield Osborn and its president Samuel H. Ordway Jr. Paul Brandwein, the foundation's education director, was also tapped to serve as one of two founding codirectors of the institute. This close relationship continued well after its partnership with the Forest Service collapsed in the late 1960s. The Conservation Foundation policy analyst William E. Shands joined the Pinchot Institute's staff in 1990, and five years later V. Alaric Sample, one of Shands's younger colleagues and a senior fellow at the foundation, became the institute's president.

Like Osborne, Ordway, and Shands, Sample routinely has framed the organization's nonpartisan stance against an otherwise caterwauling capital. In 2011, amid the then-bruising battles over the federal debt ceiling, he reflected that "no single political philosophy has a monopoly on wisdom or truth, and no one interest in society is infallible. It is essential in our democracy that there be a healthy competition of ideas through processes that even the Founding Fathers themselves knew would be messy, sometimes agonizingly slow and, when necessary, self-correcting."[6] He was convinced that because these principles guided the Pinchot Institute, it had been able to avoid the ideological pitfalls that trapped other policy entities.

The Pinchot family has kept the institute grounded in another sense, and *Seeking the Greatest Good* recounts its three-generations-long connection to Grey Towers and the institute. Gifford Bryce Pinchot's service began with his conception of the Pinchot Institute at Grey Towers as dedicated to advancing conservation in the postwar years, a proposition that the Forest Service accepted and worked with the Conservation Foundation to achieve. As a member of the institute's inaugural board of directors, and until his death in 1989, Pinchot was also the institute's voice of conscience. He nudged and cajoled the Forest Service to remain true to its commitment to the institute, and when it did not he was quick to challenge its actions. Pinchot's sons, Gifford III and Peter, and their families, have extended this connection to the Forest Service, Grey Towers, and the institute as consultants, critics, and collaborators.

As indicated in this book's title, the pursuit of the greatest good is as hopeful as it is unending. Each generation has (and must) act on terms of its own devising to determine how to live within natural systems without destroying them, a challenge that is accelerating with the warming of the planet. This underlying theme of the need for mutability, which is acutely felt in an era of climate disruption, is not new. In his dedicatory speech at Grey Towers, President Kennedy made the same claim when he asserted that Gifford Pinchot's contribution "will be lost if we only honor him in memory. It is far more fitting and proper that we dedicate this Institute as a living memorial," an approach ensuring the institute would evolve as it looked "to the future instead of the past."[7]

The most recent expression of this living legacy was on display on that warm summer day as Grey Towers' staff and the directors of the Pinchot Institute planted willows and dogwoods in the Paulins Kill floodplain—an act of remembrance and optimism.

A LIVING MEMORIAL

CHAPTER ONE

This Old House

You must know how deeply I feel about our town. For most of my life the grey house back on the hill has been my home, and it will always be my home. From its doors I can see our town spread among the trees in real beauty. And many times, I must confess, I let all my duties collect dust while I stand up there, looking at Milford and thinking of the many men and women who have lived in the valley and made our town.

—Governor Gifford Pinchot, 1933

They came on a pilgrimage. In September 1961 the Gifford Pinchot Chapter of the Society of American Foresters held its annual meeting in Milford, Pennsylvania. It was the hometown of the group's namesake, who had established the national society to which they belonged, had been the founding chief of the U.S. Forest Service, and later served two terms as governor of Pennsylvania. Most of all they came to commemorate Gifford Pinchot's death fifteen years earlier by visiting his grave.

But it was not the great one's neoclassical mausoleum that caught their eye, even though it sits gracefully on a small hill within the cemetery that Frederick Law Olmsted had designed and that Gifford's parents, James and Mary Eno Pinchot, had helped underwrite for the community. No, they were most struck by the gravesite's naturalistic setting and the stipulations about it that the legendary forester had left in his will. A copse of pine seedlings was to be planted around the crypt and the whole was to be left undisturbed—no mowing of the grasses or pulling of any weeds that might take root, no pruning of the trees as they matured, no manicuring whatsoever. As the foresters crowded around the quiet, well-shaded spot that late summer, they were initially puzzled by Pinchot's decision to let nature have its way with his final resting place: "For a while, the local people say there were complaints, especially from visitors, that it looked unkempt and forgotten." Then they began to see the landscape as Pinchot had intended it to be seen. After a decade and a half of growth, one of the visiting foresters noted, "the seedlings have pushed up and shaded out the grass and weeds, and the area is taking on a 'forestry' look, which the Governor probably had in mind when he made his will."[1]

Taking the hint, these foresters, few of whom would have met, voted for, or worked with Pinchot, started discussing how best to memorialize his service as the Keystone State's chief executive, his earlier transformative work as its commissioner of forestry, and his many other contributions to the nation—to the land itself. The conversation amplified as they toured the village that at that point had been home to six generations of Pinchots, spent time wandering around the family's estate, Grey Towers, a bluestone chateau for which Richard Morris Hunt had served as lead architect, and trekked up the slope behind the mansion to the former site of the Yale Forest School's summer camp, an experimental forest on the family's property. As they chatted over meals and coffee breaks, they resolved that the Society of American Foresters should give thought "to the possibility of having the Gifford Pinchot home and estate, which consist of numerous plantings, made into a national forestry shrine, as the cradle of American forestry."[2]

Thinking along the same lines was Gifford Bryce Pinchot, the only child of Gifford and Cornelia Bryce Pinchot. He had inherited the property when his activist mother—a feminist, a three-time candidate for Congress, and a globe-trotting Goodwill Ambassador for President Harry Truman—died on September 9, 1960. Pinchot, a biochemist and medical researcher at Johns Hopkins University, loved the old place, knew its nooks and crannies, and had tramped every inch of its rolling terrain. Best of all was the Sawkill Creek, a tumbling rush that flowed through the estate; by its waterfalls and deep pools, his parents, avid anglers both, tried to teach him the art, craft, and joy of fishing. The first lesson did not take, or so his famous forester father wrote with comic cast: "When my son announced his participation in the affairs of this world with a barbaric yawp of infancy, his Mother and I destined him to be a fisherman. Anxiously we waited for the time when he might take his first trout, and take an interest in taking it." The keyed-up parents did not wait overlong; born in 1915, young Gifford was quickly, if metaphorically, thrown into the deep end: "At age three, accordingly, we explained to him about fishing, which dissertation he obviously failed to comprehend, and asked him if he didn't want to catch a tiny speckled little trout." The trio headed out of Grey Towers, slipped past Amos and Ruth Pinchot's Forester's Cottage a couple of hundred yards to the south, on their way to the Sawkill; "the cortege moved in solemn procession to the stream. It was no light matter. The son and heir was about to begin his career—catch his first fish." Alas, when his father snagged a trout, and handed the line to his toddler to reel it in, the boy affected little interest in the grand event. "Was it possible that the son of such parents could fail to love to fish? We couldn't believe it, and, what was more, we didn't intend to stand it." They did the only thing they could think of, letting nature take its course: "because we did not press him, before he was ten Giff was casting a workmanlike fly"; he was hooked for life.[3]

This youthful connection only complicated Gifford Bryce's later inheritance: letting go of the estate would mean stepping away from some of his beloved memories, separating himself from his affective patrimony. Yet for all of its attractions, Grey Towers was a sink. Built in 1885 on a promontory high above Milford and the Delaware River, the bluestone manse, patterned after Lafayette's LaGrange chateau in Courpalay, France, was badly in need of repair. The funds required to rehab its twenty-one thousand square feet of floor space, vast slate roofline, hard-rock foundation, and the interior warren of rooms and salons would have put a major strain on the family's resources, especially when combined with local property taxes. Cornelia's death marked the end of "a grand way of living," her grandson Peter Pinchot remembered; the mansion "clearly did not fit the scale of living of the remaining Pinchots." It did not match up with Gifford Bryce's abiding passion for sailing, either (which he also gained from his parents). His "great love was the sea," his wife, Sarah (Sally) H. Pinchot once said; "this was the problem, this was why Grey Towers came to the Forest Service—we couldn't seem to rid ourselves of the desire to get in a boat and go as far as we could." To sell his heritage, however, troubled Gifford Bryce, especially if it led to the property being subdivided for suburban housing, a process then chewing up wide swaths of the New York metropolitan area. Having grown up in a household of two dedicated conservationists, he could not imagine unleashing the bulldozer on lands that members of the family had owned since the early nineteenth century. That's why he "spent two years agonizing about what to do with Grey Towers."[4]

The Salvation Army offered a possible way out. Having read of Cornelia Pinchot's demise, Lt. Colonel Roy S. Barber, the organization's national welfare consultant—which meant he served as outreach and liaison coordinator between the army and a host of national and international institutions—wrote Pinchot a letter of condolence that came with an offer. The

Salvation Army would be very interested in securing Grey Towers, erecting there a memorial to Gifford Pinchot's conservationist career, and using other portions of the estate for youth activities. Pinchot was intrigued. Acknowledging he had not made up his mind about Grey Towers, and indicating that he and his wife would continue to reflect on their options over the coming months, he promised to have a clearer sense of what to do by the subsequent summer, a delay that fit with Barber's long-term fundraising plans. But Pinchot also cautioned Barber that only certain uses could be sustained on the site: "my aunt, Mrs. Amos (Ruth) Pinchot, lives quite close to the main house . . . [and] for this reason a summer camp for children on the property . . . would certainly be out of the question and I would not feel free to sell you the property if I thought it would be used for this purpose."[5]

This interchange is revelatory. Barber's suggestion that Grey Towers could be turned into a memorial gave Pinchot the idea of another possible owner, the U.S. Forest Service. At the same time, in identifying his aunt's stake in Grey Towers, he pinpointed what he and the federal agency would also learn—that the interlocking lands the Pinchots owned separately and collectively would have to be factored into any decision it made about the site's future management. "I'm sure this whole thing is very difficult for you to consider as it has been for us," he wrote his aunt Ruth in June 1961: "We've been thinking about it pretty consistently since mother's death and there doesn't seem to be any way that we can see to keep it for our own use. It's been no fun to reach this conclusion but we do feel quite sure at least that it is the right one," adding as an underscore, "I'm awfully sorry to bring this difficult problem to you but there just doesn't seem to be any alternative."[6]

Before he could act, he had to protect. Over that summer and fall, Pinchot spent thousands of dollars to repair roofs, gutters, and entryways for the main house and some of its outbuildings as a minimal investment in their maintenance, "before the old place collapses." As the architects, landscapers, carpenters, roofers, and painters came and went, Pinchot concluded that the Forest Service might be his best option. In retrospect the decision seems obvious. Because of his father's status as agency founder, because his parents had used Grey Towers as a mini-Chautauqua, hosting labor organizers and industrialists, artists, actors, and writers, as well as conservationists and conservatives, feminists, Old Guard politicians, and young radicals to debate contemporary issues day and night, indoors and out, would not that same ferment be nurtured if Grey Towers became a center devoted to social inquiry? Given the family's storied association with the Forest Service, there may have been no better organization to steward the estate, bringing it back into active engagement with the body politic.[7]

These resolutions led Gifford Bryce straight to his typewriter in September 1961 to write to Richard E. McArdle, chief of the Forest Service: "My wife and I have been living at my father's place in Milford, Pennsylvania, for part of this summer. It occurred to us that the Forest Service might possibly be interested in having it and I have accordingly investigated the possibilities of giving it away." Noting that the property was in a trust, "and the whole situation is rather complicated"—his Philadelphia lawyer Harold B. Bornemann later indicated that to "get the property out of the trust and the title into the United States was one of the most difficult . . . assignments I ever tackled"—still Pinchot was convinced that a donation was possible. He hoped too that the concept would be alluring: "The setting is quite beautiful with gardens, pools and so on, and the whole place might conceivably be used as a museum. Another possible use would be as a conference center for Forest Service or Department of Agriculture personnel, something along the line of Harriman's old place in Harriman, NY."[8]

The idea caught the agency's attention: "You suggest an extremely intriguing possibility," responded Associate Chief Arthur W. Greeley, and he hastened to take Pinchot up on his offer for a tour of the mansion and its grounds. The one initial catch he voiced was significant: "Our problem is to come up with a fully justifiable way by which we would make use of it in a manner that is in the public interest." That hesitation remained in the aftermath of the agency officials' visit that October. "Having lived and worked in the traditions set by your father," wrote Hamilton Pyles, the regional forester for the agency's eastern region, he was particularly struck by the revered man's almost palpable presence: "The stamp of G. P. as we knew him or thought of him is certainly left in 'his' room on the second floor. I sincerely hope if nothing else happens that this arrangement of personal things can be preserved intact in the most appropriate place." Whether that would occur under the agency's care was as yet unclear. "Your generous proposal regardless of outcome is going to be deeply appreciated by all of us in the Forest Service—not only those who knew G. P. but the full rank and file."[9]

By December the agency was considerably less guarded: sometime that fall, the Conservation Foundation, a New York–based think tank devoted to the protection of the nation's natural resources through research and education, had approached it with a proposal to cosponsor a conservation center. The notion gained immediate credibility within the agency, given that the foundation's board of directors contained a who's who of the conservation movement. Its founder was Fairfield Osborne, then-president of the New York Zoological Society and author of the provocative best seller *Our Plundered Planet,* and its president was Samuel H. Ordway Jr., an attorney and author; Laurence Rockefeller, the financier David Hunter McAlpin, and the playwright and actor George Brewer were among those who promoted the foundation's agenda. The idea was well received too because these individuals were also well connected to Forest Service leadership through allied organizations, particularly the Natural Resources Council of America (1947–1981), which functioned as a clearinghouse for most major national and regional groups focused on conservation; it may have been under its aegis that the first discussions about the conservation center occurred. The "lack of an existing facility, or funds to develop a new one, represented critical obstacles to such a joint project," Greeley advised Pinchot in late December. "Your offer of Grey Towers and adjoining property would now appear to make such a center possible," so much so that the two entities already had agreed on a working name—the Gifford Pinchot National Conservation Center.[10]

They had also agreed on a preliminary sketch of its future work. The facility would host gatherings of leading organizations to rethink the nation's natural-resource policies, offer leadership training for youth and adults, civic and school groups, and for Forest Service personnel, and establish a Gifford Pinchot museum for visitors to Grey Towers and a possible nature museum featuring a demonstration forest. Central too would be the development of a "national center for the preparation and testing of school curriculum materials for teaching conservation," for this is "a badly neglected element in conservation progress to date." Noting that "one of your father's goals was an American public aware of their stake in the conservation of our vital natural resources," Greely predicted that a "memorial center devoted to this purpose could become the force for a much-needed public education in conservation," and its impact would be ramified by Grey Tower's location near New York City, presenting "an opportunity to reach a large and vital segment of our population."[11]

For all the enthusiasm and flattery that percolated through Greeley's letter—"we feel that no more fitting memorial to your father and his historic accomplishments could be established," that this center "could well be the same sort of dream that marked his trail blazing in conservation"—for all the Conservation Foundation's apparent eagerness to cooperate on this venture, and for all the Pinchot family's embrace of the prospect, there were some distinct challenges. The cost of repairing Grey Towers was substantial, according to Forest Service engineers, though the agency hoped that the "participation of the Conservation Foundation . . . would solve this financial aspect." Uncertain too was whether the small amount of adjoining property that Pinchot had proposed donating—fifty to one hundred acres—would be sufficient for the center's then-expansive mission. Believing that it might not, the Forest Service initially proposed that the Pinchot family transfer all its property to the U.S. government. Not yet determined was which of the two entities would be directly responsible for day-to-day operations, maintenance, and funding. Creating a public-private partnership between the federal agency, non-profit foundation, and family, the first of its kind for any of the collaborators, was going to be a difficult task.[12]

It would take eighteen months before the matter was settled. That time was consumed with round after round of negotiations between the three parties' lawyers and in face-to-face meetings between the Forest Service's recently formed education office and its analog at the Conservation Foundation. Every bit as complicated was the internal discussions within the Pinchot family about the amount of land to donate and under what conditions and what protections would be implemented to ensure Ruth Pinchot's and her heirs' privacy and viewshed. Her concerns led to a slow shrinking of the amount of land under discussion, from 300 acres to the final donation of 101 acres. A memorandum of understanding then went through numerous drafts, but its basic gist was that Gifford Bryce Pinchot would "sell" the property to the Conservation Foundation in exchange for ten noninterest-bearing notes of $25,000 each, with one to mature every year for the next ten years (but which Pinchot had agreed to forgive). The foundation in turn would deed Grey Towers and the adjoining acreage to the U.S. government, at which point the two organizations would jointly operate the facility.

The transfer plan appeared reasonably straightforward, yet the foundation started getting cold feet, perhaps an early sign of some of the troubles that five years later would end its commitment to the fledgling institute. In March 1962, Samuel H. Ordway Jr., as foundation president, "raised some question as to a guarantee that Dr. Pinchot will not suddenly change his mind and demand payment of the notes." Once that fear was allayed, another arose the next month: the foundation decided it would not deed over the whole property to the government immediately, so long as its notes to Pinchot remained outstanding, a full transfer that might have required ten years to complete. At the bottom of the letter in which his attorney notified him of this latest snafu, Pinchot scribbled "Please—what is on Ordway's mind"?[13]

The Forest Service was wondering the same thing. Learning that the foundation was in some doubt about the Forest Service's long-term commitment to Grey Towers—surely a projection of its own anxieties about the collaborative arrangement—Chief Edward Cliff tried to reassure Ordway of the agency's enduring engagement: "As to the Forest Service's intentions for the full future development of the Institute, let me assure you that we are, if possible, even more enthusiastic about the proposal than The Foundation, and foresee our participation in the Pinchot program into perpetuity." But because Cliff and his staff were convinced that the "outdoor education potential of the property as an area for demonstrations of land type and land use, as a center for school programs and institutes for teachers, equals the potential of the house as a conference center," the agency would not "accept less than the original intention of Dr. Pinchot to have the entire property deeded to the Forest Service." As an olive branch, Cliff suggested that the foundation insert in its deed of conveyance a binding reservation that "would assure it of a part and voice in the overall planning, development and operation of the program of the Institute." If this compromise was acceptable, then it would allow the two organizations to "reach our September target date for beginning of the Institute['s] operation."[14]

His proposal broke the deadlock, even though it took another eight months to finalize. "In view of Mr. Cliff's letter I can't foresee any difficulties that are likely to arise. I can't tell you how pleased I am with these developments," foundation board member George Brewer reported to Pinchot in May, and "the creation of the Institute and the prosecution of its programs can, and I feel sure will be, of incalculable value to the conservation movement which your father initiated more than half a century ago."[15]

Over the next summer and fall, Ruth Pinchot and her children, Antoinette Pinchot Bradlee and Mary Pinchot Meyer, and her stepson Gifford Pinchot,[16] signed over their rights to jointly owned property; the family's matriarch also received carefully negotiated assurances from the Forest Service about "maintaining the privacy of her woods." To reflect her assent, a new name for the program at Grey Towers was coined—the Pinchot

Institute for Conservation, so that "recognition of both Gifford and Amos Pinchot will be given." After nonstop tinkering, a final draft of the memorandum of understanding was hammered out and finally signed in late December; one year after the idea of the institute had been formulated it was made legal.[17]

Milford was agog. Recognizing the economic potential the Pinchot Institute might generate locally, business leaders formed the Pike County Chamber of Commerce; its first act was to establish a special committee "to be of any assistance in furthering the idea of Milford as the Cradle of American Forestry." In the interlocking network of such commercial ventures, the chamber's secretary Norman Lehde, whom Gifford Bryce Pinchot knew from elementary school and who recently had stepped down as the village's mayor, used his editorial platform as publisher of the weekly *Pike County Dispatch* to tout the struggling region's good fortune. Setting the news in a wider context—"there seems to a general inclination of the Nation and States to build memorial parks to noted persons, or purchase old homesteads that have become famous. We have Mount Vernon for Washington, Monticello for Jefferson and one for Monroe, and many state shrines to honor former notables"—he argued that "the famous Gifford Pinchot place" belonged in that distinguished pantheon: it has a "long and notable reputation as the show place of this part of the state and especially Pike County." Were the then-current negotiations over the creation of a national conservation center at Grey Towers to fail, Lehde had a backup plan; he urged the state to turn it into "a public memorial park and museum." Either way, the timing was propitious: with "the big Tock's Island Dam and Reservoir getting nearer," a reference to the Army Corps of Engineers' mega-project slated for the Delaware River just south of Milford (which ultimately was not built and instead became the Delaware Water Gap National Recreation Area), the Pinchot memorial would help lure "thousands of visitors . . . yearly." Through his largesse, Gifford Bryce Pinchot was building on the philanthropic legacy of previous generations of his family in aid of Milford and its environs.[18]

Predictions of the center's economic impact escalated following a May 1963 press conference in which the Forest Service laid out its plans for Grey Towers. The regional forester Richard Droege, the director of the Office of Information and Education Clint Davis, and Matthew Brennan, then-chief of conservation education and soon-to-be the agency's on-site director of the Pinchot Institute, indicated that upward of ten new staff would be required to produce the expected curricular programming, launch the center's publication efforts, and convene and manage the many planned national, regional, and local conferences; they let slip that neither these employees nor the institute's stream of guests would be housed or fed at Grey Towers, a revelation that sparked gleeful speculation of a local real estate boom and a much-needed boost to area hotels and motels, diners, and restaurants. That local cash registers would ring up one sale after another was not far from publisher Lehde's mind when he cheered the Pinchots' "act of conservation," their gift of Grey Towers to the American people that would provide a "tradition inspired locale . . . for the men who carry conservation into the atomic age."[19]

Then arrived the news that President John F. Kennedy was going to attend the dedicatory celebrations slated for September 24, 1963. From Grey Towers and Milford, the chief executive would kick off a "conservation caravan," carrying him on an estimated hundred-thousand-mile journey across the United States, during which he would affirm his environment commitments in the run up to the 1964 election. "When Pres. Kennedy and the other dignitaries, including Gov. Scranton and Sec. of Agriculture Freeman . . . ascend the platform," Norman Ledhe predicted, "there would be many local people in the crowd who will feel a special elation and joy." Like William Hinkel, Grey Tower's caretaker, or retired state trooper John Frank, the family's onetime bodyguard, like those who were on a first-name basis with Gifford and Cornelia or who simply had heard "what a great fisherman the governor had been," many more would be present who seventeen years earlier had stood "with bowed heads when the funeral services were held for the forester at Grey Towers." Bound together by a weave of memory and affection, Lehde believed that the charismatic president and ordinary citizens on September 24 would bear witness to what he called "a moment of fulfillment," a secular (if mawkish) second coming, for on that momentous day Gifford Pinchot "will live again as he and his family name is honored by the nation he served."[20]

CHAPTER TWO

September 24, 1963

The dedication promises to be just about the biggest and most exciting day in the history of Milford where Gifford Pinchot's great-grandfather, Constantine Pinchot, settled and opened his store in 1819.

—*Port Jervis Union-Gazette,* **September 6, 1963**

JFK dropped out of the sky. Ferried from the Stewart Air Force Base in Newburgh, New York, on Marine One, the presidential helicopter, he put down at a makeshift landing pad at Grey Towers. The president's stay was brief; one reporter timed his visit from touchdown to take off at exactly seventy minutes. Yet in that short period he toured the Pinchot family's ancestral home, pressed the flesh up and down the rope line, paid his respects to family matriarch Ruth Pinchot in her adjacent abode, Forester's Cottage, and delivered the last of five speeches dedicating the new Pinchot Institute for Conservation Studies—a jammed-packed schedule that required split-second timing and precise preparation.

An advance team from the Secret Service had been in Milford for more than three weeks, mapping out the president's future movements from the moment he landed to the second he departed. They repeatedly surveyed the tree-shaded, grass-sloped amphitheater where the public ceremony would occur, an outdoor space that Cornelia Pinchot had constructed in the 1930s so that she and her husband could hold political rallies on behalf of their tireless electioneering; there they also hosted an annual ice-cream social for the village, sweet occasions that reminded the family and local residents who they were in relation to one another. After carpenters knocked together a platform to hold the White House Press Corps' television cameras, the federal agents had them move it around the site until they were satisfied that it would not hinder their protection of the president. In an eerie fore-shadowing of the tragedy in Dallas that would occur two months later, the Secret Service was also dissatisfied with the dais from which the president and other dignitaries would speak. Because it had no crawl space under-neath, the agents required carpenters to stand by so that at the last minute they could pry up select boards and then, after a careful examination, nail them back down just before the ceremonies commenced. The Secret Service also posted an officer "at the stand to prevent tampering or the introduction of some kind of infernal machine that could endanger the President's life."[1]

While local and state police coordinated their strategies for controlling the flow of an expected crowd of ten to fifteen thousand into and out of Milford, the Forest Service was rehabbing Grey Towers' plumbing, wiring, and ventilating systems, converting portions of the building into offices and conference rooms and widening roadway access, all at a feverish pace that accelerated as September 24 neared. It set up an off-site office too, squirreled away in a local motel. The crowded, smoke-filled suite, with phones ringing nonstop, had, one of its denizens remembered, "a dresser lined with liquor bottles"; it was a well-stocked command center that op-erated 24/7. The agency moved a lot of earth to construct the one-day he-liport, organized the construction of stages and the stringing of ropes and cordons, and contractors unspooled and buried underground an estimated three miles of cables for teletype machines, telephones (three of which were presidential hot lines), and telephoto circuits to ensure that the White House and the press could tap into a global communications network.

The landscape itself was reconstructed: Forest Service crews cleared away and burned large piles of brush and clippings, creating a smoky plume so large that downwind it "dirtied some Monday morning laundry." In the last days, more than sixty-five employees were detailed to Milford to serve as ushers during the celebrations. The essential obligation was that each individual had to have an official Forest Service uniform, a require-ment that dismayed the future chief Dale Robinson, who had been tapped to serve but for lack of official regalia could not participate. Those who did never forgot the crazy, "mad whirl" of energy they expended ensuring that the event came off without a hitch.[2]

"Serene Milford," one newspaper described the community that was anything but. City workers cleaned streets and sidewalks, the garden club replanted window boxes along Harford and Main Streets, the central arter-ies at whose intersection the Pinchot's nineteenth-century store had stood facing the family's old homestead. Volunteers spruced up the parks and hung Old Glory and red, white, and blue bunting from every light stanchion, telephone pole, and tree; the Delaware Valley High School Band, slated to entertain the crowd before the president's speech, practiced nonstop. As for the horde of guests expected to rush into town, a committee was formed so that they would be "made welcome and rendered service in a scene that might rival a Cecil B. DeMille production."[3]

With the pulsating whomp-whomp-whomp of a pair of Sikorsky helicopters, the show began. As first one and then the other touched down, the presidential entourage, including Agriculture Secretary Orville L. Freeman, Secretary of the Interior Stewart Udall, and sisters Antoinette Pinchot Bradlee and Mary Pinchot Meyer—daughters of Ruth Pinchot and friends of the Kennedys—disembarked. Uniformed Forest Service employees guided them to a caravan of limousines for the short ride uphill to Grey Towers. For two young Pinchot relations, the president's arrival was memorable, if for different reasons. Although she was only three at the time, Bibi Gaston, great-granddaughter of Amos Pinchot, recalls "the helicopter landing and the cheering crowds," remembering too the "electricity in the air, which, my adult self imagines was probably the case almost everywhere JFK appeared. Grey Towers that day was Camelot." What most struck her older cousin Peter Pinchot, son of Gifford Bryce and Sally Pinchot, "was watching the president come walking down steps of the helicopter under the spinning rotors which were blowing up quite a breeze. And not one hair moved. I thought: 'What does this man do to his hair?'" The well-sprayed coif was not what caught his mother's eye—she and her husband had guided the president through Grey Towers and later accompanied him while he visited with Ruth Pinchot. She was captivated: Kennedy "was a charming individual, whatever you might have thought of his politics; he was a charmer alright."[4]

The crowd was waiting. Enveloped by Secret Service agents, Kennedy made his way down to the amphitheater, screened by the landscaping and his guards. Only as he neared the open, gently terraced space did some in the audience get a glimpse of the president. In an instant the air resounded with "the noise of applauding adults, cheering boys and girls." Someone shouted "there he is!" and that was Forest Service Chief Edward Cliff's clue: he stopped introducing the podium guests, stepping away from the microphone as the band struck up "Hail to the Chief." Amid the thunderous reception, journalists still managed to pay close attention as Kennedy shook hands with those assembled on the stage, waiting for the president and Governor William Scranton, who was to deliver a brief welcome, to greet one another. Their cordiality was unforced, though one reporter speculated that each man knew "full well they could easily be opposing candidates next year in the presidential election."[5]

But on this day everyone was there to see and hear the president. That's why small children were hoisted up on their parents' shoulders, why a grandmother sat in a chair under an upslope apple tree, and why folks of all ages, who formed the "sea of faces" arrayed before the platform, probably did not attend as carefully to the three speeches that preceded Kennedy's address. But as Gifford Bryce Pinchot, Samuel H. Ordway Jr., president of the Conservation Foundation, and Orville Freeman "eased up to the microphones to speak," their words mattered. Each was a stand-in for the three entities without which there would have been no ceremony that day. The Pinchot family was donating Grey Towers to the American people and the Conservation Foundation and the Forest Service, through an unusual collaborative partnership, were committed to bringing the Pinchot Institute to life, with a mission to advance through education the nation's appreciation for and understanding of how to steward its natural resources and protect its imperiled environment. Each story came with a past, a history that opens up the larger forces—familial, ecological, institutional, and political—that converged in ways fortuitous and fateful and that led these four men on a cool and crisp September afternoon to stand before the animated crowd sweeping up the hill.[6]

CHAPTER THREE

Home Grounds

A huge towered Camelot set on the side of a treeless stony hill, with the usual French dislike of shade, inherited from a Gallic ancestor.

—Cornelia Bryce Pinchot

Gazing out over the boisterous crowd of family, friends, and luminaries, Gifford Bryce Pinchot was reminded of similar gatherings that had occurred whenever one of his parents hit the campaign trail, which was often: "This seems to me to be a continuation of the wonderful days when my father and mother lived here, and I can only think how much they would have enjoyed being here to welcome you themselves." Grateful that President Kennedy was on hand to dedicate the Pinchot Institute for Conservation Studies, he was convinced too that "my father and mother would feel the same way as we do, that this is the perfect use for Grey Towers. Conservation and the U.S. Forest Service were the concept and the organization nearest my father's heart. There could be no more fitting use for his house."[1] The personal had become public; the familial had become national.

In so framing the transfer of Grey Towers to its new owners, Gifford Bryce spoke in language that was itself directly linked to the Pinchot family's notion of and commitment to civic virtue and democratic obligation; these concepts drew on a welter of French and American cultural legacies that stretched back 150 years to when the first Pinchots arrived in Milford in the early nineteenth century from France in the tumultuous aftermath of Napoleon's defeat at Waterloo. They had lived in Breteuil-sur-Noye, about sixty miles north of Paris where Gifford Bryce Pinchot's great-great-grandfather Constantine was a prosperous and politically engaged merchant; he and his son Cyrille Constantine Désiré were reportedly thrilled when Napoleon broke out of exile, joy that collapsed after Wellington destroyed the French army. Fearful that they would suffer under the Bourbon Restoration, Constantine, his wife Maria, and Cyrille decamped, fleeing to New York.[2]

They came with enough financial resources for Constantine to resurrect his business interests in the booming city of the new republic. He achieved some success too, enough at any rate that in 1819 he sold out, using the profits to purchase eight hundred acres of meadows and woodlands surrounding Milford, Pennsylvania, and a lot in the town on which he built a home and store. He had selected Milford in part because of its geographical location and natural resources: situated in the northeastern corner of the state and at the head of what is now called the Delaware Water Gap, there was a still-ample supply of timber, which, once harvested, could be rafted south to market. The surrounding area had also drawn other French (Protestant) émigrés; one of the earliest had been the man of letters Guillaume-Michel Saint-John de Crevecoeur: his utopic depiction of agrarian virtue, of his neighbors whose "simple cultivation of the soil purifies them," played well in the salons of Paris and Philadelphia, but they were of less account along the mud-choked roads of Milford. Certainly that was true of the Pinchots, for whom the region's rough lanes and fast-flowing rivers and creeks were the keys to their economic future. These diligent republicans were on the make.[3]

Foremost of the family's entrepreneurial projects was the store, sited on the crossroads that demarcated Milford as the political hub and commercial center of Pike County. Reinforcing its centrality was the ambitious town's streetscape, which the Circuit Judge John Biddis had platted in the late eighteenth century after Philadelphia's interlocking grid of streets and alleys, a circulatory pattern that proved highly advantageous to the newly arrived merchants. Because the community was also located along interior trade routes that intersected with the riparian movement of goods and services between the agricultural frontier and bustling New York and Philadelphia, the Pinchot's store became a pivotal exchange. Through its doors flowed local garden-fresh produce, tools, and other necessities, as well as finished goods—cloth, linens, and tobacco—from urban manufactories. With some of the monetary gain from this lucrative trade, the Pinchots purchased additional wood lots and arable land, hiring tenant farmers to clear and work them. When Constantine died in 1826, he was among the county's largest landowners.[4]

His son Cyrille built on this legacy. While his mother Maria (and later his wife, Eliza) ran the family store, he pursued land-speculation schemes across Pennsylvania and New York State and farther west in Michigan and Wisconsin. Particularly interested in forested lands, like his contemporaries from Down East to the Great Lakes, he bought up woodlands, cleared them during the winter, built rafts out of sawlogs and boards, and then, as the snow melted, ran them downstream to sell in river ports along the Delaware. With the returns, Cyrille Pinchot invested in additional timber stands, repeating the cycle.[5]

This generation of farmers and loggers produced material riches and did considerable environmental damage. With only market demand to regulate their actions, these entrepreneurs sliced through the American wilderness, sending millions of board feet of pine, oak, maple, and hemlock to mills; once harvested, they pulled up stakes for the next cut, leaving behind denuded hills, eroded terrain, silted rivers, and fire-prone terrain. Peak production in Pike County occurred between 1830 and 1860, when "sawmills dotted every mountain stream; lumber, manufactured and in the log, covered the banks wherever an eddy could be found suitable for rafting, and in the spring and fall a majority of the male population were floating their hard earned products down the Delaware in search of a market." After the Civil War, observed James Elliot Defenbaugh, the county's onetime "dense forest of white and yellow pine, oak, ash, and hickory" had been slicked off. In time, members of the Pinchot family—notably James, his son Gifford, and their descendants—would work to control industrial production in the woods, but advocating for such restraint was not Cyrille's calling.[6]

Indeed, for him, the machine in the garden was a constructive engine. It enabled him to harness nature's energies to enhance his material world (and not incidentally that of his progeny) and to contribute to the economic growth of the republic. To expand Milford's commercial prospects, he was involved in the development of stage lines and the Milford and Owego turnpike (whose route would roll past the property on which Grey Towers would be constructed); he worked assiduously to bring the railroad to Milford, knowing that it would otherwise redirect the regional flow of goods and services away from his ambitious little burg. Despite Pinchot's best efforts, which included securing state legislative intervention that forced the Erie Railroad to build a spur-line bridge across the river (which it never completed, a span to nowhere), the tracks, warehouses, stations, and yards remained on the east bank of the Delaware, boosting the prospects of Port

Jervis, New York. Still, by the 1850s Cyrille Pinchot had become the largest taxpayer in Milford, and he and his family inhabited a stately Greek revival house across the street from the Pinchot store. With Cyrille, private good and public service had converged.[7]

For all his successes, the rising generation sought greater opportunities elsewhere. Among Milford's motivated sons and daughters who left town were Edgar, James, and Mary Pinchot. Mary headed east to Bridgeport, Connecticut, with her husband. James, like Edgar, took the Erie Railroad south to New York City where the brothers made their fortunes in that city's explosive pre–Civil War economy. James's financial success from the manufacture and sale of wallpaper and other decorative furnishings was augmented by his marriage to Mary Eno in May 1864. Smart, shrewd, and driven, Mary was the well-connected daughter of Richard Amos Eno, one of New York's wealthy merchants and real estate developers. Mary's dowry and the familial connections that she and her parents enjoyed added immeasurably to the young couple's start in wedded life. So advantageous was it that James was slowly able to pull back from the direct running of his business affairs after his son Gifford's birth in 1865. "James was very lucky in that he came across the concept of Enough," observes his great-grandson Gifford Pinchot III; "he realized that he had made enough money and he didn't need to do that anymore and could turn himself to another purpose in his life—this life of the mind and what was needed in society." Among those retirement projects to which he dedicated some of his riches was the renovation of Milford, transforming the declining entrepôt into a tranquil tourist resort.[8]

Starting with his 1863 razing of a building next to the Pinchot family's store for a new post office, James began gentrifying the dusty village of his youth. Deputizing his father to purchase additional town sites, from New York City James planned a new library and chapel modeled loosely after the ivy-covered structures that graced the rural English communities he had toured in 1871: "I wish everyone in Milford could see [these hamlets] to know how much could be done in beautifying our village," he wrote to his mother. These buildings offered a moral perfection and social uplift that he believed would benefit the rougher inhabitants of Milford, whom he compared to the Irish peasantry he encountered while visiting the Emerald Isle. James Pinchot was convinced that a gentle landscape gave birth to a genteel people, a refined environment in which to raise his progeny.[9]

Pinchot was not unique in his hunger for the refinement money and status could bring. As with other members of the American cultural elite, he built these concerns into the very architecture of his country estate in Milford, construction for which began in the mid-1880s when his oldest son Gifford was twenty. He hired the celebrated architect (and close friend) Richard Morris Hunt, whose designs catered to the aspirations of the upper crust. Hunt housed the elite in sumptuous abodes in New York City's high-end residential neighborhoods, built their summer "cottages" in Newport, Rhode Island, and notably was the architect for George W. Vanderbilt's palatial manse the Biltmore Estate in Asheville, North Carolina. It was on Vanderbilt's vast forests surrounding his estate there that Gifford Pinchot would launch his forestry career in the early 1890s.

For the Pinchot family's home, which they came to call Grey Towers, Hunt constructed a Norman-Breton bluestone manor that dominated the physical and social landscape. An imposing, fortress-like exterior, complete with three sixty-feet-high turrets, was matched by an interior that contained a great entrance hall, twenty-three fireplaces, and forty-four rooms, each crammed with furnishings matching those "of the old baronial days." The manse's siting on a "commanding eminence," overlooking the village of Milford and the Delaware River, intensified its visual impact: the sheer size and scale of this "summer castle," when combined with its self-conscious evocation of the Pinchots' French ancestry (a bust of Lafayette, tucked in a niche in its eastern wall, gazes across the rolling hills of western New Jersey toward France), draws all eyes uphill.[10]

Over the years, Mary Pinchot would urge her husband to elevate his sights even more: "I believe you capable of a higher development than you will get if tied down to paper hangings," she counseled in the 1880s; "I think it a mistake that a man of such noble aspirations and large capacity should not fill a larger sphere." It is noteworthy that she offered this particular advice when the Pinchot and Eno families were involved in an internal debate about the future of young man Gifford. Impressed with his Yale-educated grandson, Amos Eno urged him to enter the Eno family's real estate business. But Mary, James, and Gifford resisted, insisting that his talents lay in public service and more precisely in the as yet unknown field of forestry. "James was actually the pivotal character who recognized the damage that had been done by our family and many other people," Peter Pinchot remarked about the choice of forestry as his grandfather's future occupation, "and that something needed to be done." Grey Towers, after all was so visible from the valley because the Pinchots had clear-cut the surrounding lands; James set about repairing this environmental damage on his land, and he and urged his firstborn son to take up a profession that would help restore the nation. This paired ambition lay behind the parents' resistance to Amos Eno's designs; their position would be even more persuasive, Mary Pinchot reasoned with her husband, if he set himself up as a model of engaged public service for his son, thereby deflecting her father's insistent prescription for Gifford's future.[11]

Shortly thereafter James did as his wife had suggested, becoming an important supporter of the fledgling American Forestry Association, established in 1875. Mary Eno Pinchot's influence on her children's life choices was profound too, signaled by the fact that Gifford, his sister Antoinette, and their younger brother, Amos, took on important public roles in the United States and England. The pressure, however, was greatest on the couple's oldest child: the education they provided for him, the social and political climate within which they raised him, and the expectations they routinely announced had the intended effect. Shortly after his twenty-fourth birthday, for instance, and thus well before he became a public servant or an elected official, Gifford Pinchot spoke to his fellow citizens of Milford as part of the ceremony commemorating the centennial of the U.S. Constitution. About the obligations of citizenship, the young man had a decided opinion: "We are trustees of a coming world," he declaimed in August 1889. But before embracing the future, he and his audience were obligated to acknowledge their debt to this particular place, Milford: "We are first of all . . . citizens of Pike County, Pennsylvania, and it is here we are to realize, if at all, the blessings of the great birthright which has descended to us from the courage, perseverance, and energy of our forefathers." To deserve that blessing, and to extend it to subsequent generations, required Pinchot and his listeners to accept this twofold obligation: "not only that we have a share in the commonwealth, but that the commonwealth has a share in us." A subsequent president would formulate his charge to the nation during his 1960 inaugural address—"Ask not what your country can do for you," John F. Kennedy had declared three years before he traveled to Milford to accept Grey Towers as a gift to the nation, "ask what you can do for your country." Gifford Pinchot had answered that same imperative seven decades earlier, affirming that the nation-state has "a right to our service, our thought, and action."[12]

This was the exacting code by which he and his wife, Cornelia Bryce Pinchot, raised Gifford Bryce Pinchot. The couple had met while working on Theodore Roosevelt's Bull Moose Party campaign in 1912. She came from a long line of civic leaders, diplomats, and entrepreneurs, was a staunch suffragette and children's and workers' rights advocate, and Roosevelt was convinced she had one of the keenest political minds in the United States. The former president attended the couple's wedding in 1914 and surely was not surprised when they delayed their honeymoon to barnstorm across Pennsylvania in support of Gifford's unsuccessful pursuit of a seat in the U.S. Senate. It was into the politically saturated household that the Pinchots' only surviving child, Gifford Bryce, came of age.[13]

On a biannual basis, one or the other of Gifford Bryce's parents was campaigning for office or on the hustings in aid of a colleague or a cause. Their frequent absences and absent-mindedness were among the difficulties that resulted from their "immensely busy" life in the public eye. When "mobs of political supporters" showed up at the governor's official residence in Harrisburg or at Grey Towers, Gifford Bryce's parents invariably pressed their son into playing the role of cute, diminutive host: "one of the things that killed him," his wife later observed, was that Gifford and Cornelia "didn't always remember the . . . names of everybody who showed up," and so they would "wish them on [their son] and there he would be floundering, not knowing who these people would be; you know, let the little boy amuse them." Even when the family went on vacations overseas, they were never far from the political arena. After Gifford Pinchot's first term as governor, they sailed aboard a 150-feet-long schooner named the *Mary Pinchot* from New York through the Panama Canal, visiting the Galápagos, Marquesas, and Tahiti. Yet the ship-to-shore radio crackled with news from home, and incoming telegrams kept the Pinchots, husband and wife, abreast of the latest political maneuvers in Harrisburg and Washington, D.C.[14]

Given these experiences, it is not shocking that Gifford Bryce's public service would take a different shape from his parents': "I didn't always see eye to eye with [my mother]," Gifford Bryce recalled, "especially when she deplored my decision not to follow in my father's footsteps into politics." His upbringing was formative nonetheless: "he was certainly influenced in his desire to do something to help the world," Gifford Bryce's wife Sally Pinchot observed years later; "I think that's part of why he went into medicine, he felt he couldn't, didn't want to do it the same way his family had done it." She might have noted too that his fascination with medicine's healing powers was also a consequence of his family's memorable 1928 Pacific voyage, during which he worked closely with the ship's doctor. "The cruise of the *Mary Pinchot* had a profound effect on my life," Gifford Bryce confirmed; "I had been brought up with the very strong feeling, which sounds very dated and almost ridiculous nowadays, that you should try your very best to do something useful with your life; to put it baldly, to try to leave the world a better place than you found it."[15]

The discussions in the early 1960s within the Pinchot family about the fate of Grey Towers regenerated Gifford Bryce's interest in what had been his parents' passion. He was, recalled his wife, "terribly interested, by this time, in the environment and right up until he died that was what he was most working on." His activism took multiple forms. In Milford, where he built a much smaller house after the disposition of Grey Towers, Gifford Bryce became engaged in the grassroots campaign to stop the construction of Tocks Island Dam. The idea for the flood-control and hydroelectric project on the Delaware River had surfaced in 1955, in the punishing aftermath of hurricane-generated flooding south of Milford. Yet opposition to the U.S. Army Corps of Engineers–developed plans to construct a 160-feet-high rock-fill and earthen dam emerged as soon as the first public hearings for it were held one year later. "Most shoreline residents were against the dam proposal from the beginning," one of the protestors commented: "We had to be; we knew the geology was bad." The Army Corps of Engineers never acknowledged the dam's critical design flaw, though other geologists did. When the federal government began buying up land, houses, and whole towns and then forcibly relocating residents, opposition exploded. Confronted with adverse geological evidence, growing political anger, and more economical options for reducing flood damage and water-supply improvements, the Delaware River Basin Commission voted in 1975 to cancel the Tocks Island project. (The land that the federal government had purchased or taken through eminent domain formed the core of the Delaware River Water Gap National Recreation Area that ever since the National Park Service has managed).[16]

Deriving its energy from the same kind of tactics as had been employed against the Tocks Island Dam project, the Milford community successfully fought against the construction of a nearby shopping mall. It would have undercut the area's rural character, degraded the town's potable water supply, and clouded nearby streams, including the Sawkill, a free-flowing creek that runs across Pinchot-owned land surrounding Grey Towers. Organizing local activists into the Pike Environmental Defenders, a group which included his son Peter and cousin Nancy Pittman, Gifford Bryce

Pinchot and his allies battled the development in the press and through political channels, ultimately carrying the day.[17]

Gifford Bryce's interests in contemporary environmental issues were national as well. In 1970 he joined with other scientists and lawyers to establish the Natural Resources Defense Council (NRDC) and served on one of its advisory boards. He may not have known it but in this too he was following in his father's footsteps. In 1909 Gifford Pinchot had established the National Conservation Association to track and analyze congressional debates over conservation measures, craft relevant bills, and lobby for more rigorous legislation at the state and federal level. Sixty years later, the NRDC acted similarly, confronting some of the most critical environmental issues on Capitol Hill, in the courts, and in the public arena. One of these was the mounting furor over the Forest Service's postwar practice of clear-cutting and its effects on the national forests. Gifford Bryce Pinchot jumped into that debate in 1973, when, with a group of academics, foresters, and NRDC colleagues, he toured the Bitterroot National Forest, site of some of the most aggressive harvesting techniques in the country. As Pinchot and his colleagues bore witness to what has come to be known as the "Oh My God" clear-cuts, he was reported to have said: "If my father had seen this, he would have cried." Gifford Bryce's shocked reaction set off a chain reaction. Some foresters lashed out at his amateur status, others sneered that he misunderstood his father's work. These attacks were met with counter-thrusts on the part of those foresters who had traveled with Pinchot and from environmental activists already convinced that industrial forestry on public lands was by definition a travesty. After reading a particularly galling editorial in *American Forests,* Gifford Bryce responded bluntly: "I believe that had [my father] lived to see them, the cutting practices on the Bitterroot would have indicated to him that the Forest Service, which he deeply loved, had abandoned [its] conservation principles under pressure from the lumber industry." Ten years after transferring Grey Towers to this same federal land-management agency, Gifford Bryce questioned whether the Forest Service still upheld the values its creative genius, his father, had espoused.[18]

In acting as he had at Bitterroot, Gifford Bryce was upholding another of his parents' beliefs. Shortly after his father's death in 1946, for instance, Cornelia Bryce Pinchot praised her husband's convictions: "Gifford Pinchot was always the first to proclaim the principle of growth, of development, and of renewal as central to the conservation idea," yet he also insisted "that conservation must be reinvigorated, revived, remanned, revitalized by each generation." Only in that way could its implications, "its urgencies, its logistics" be translated by the present for the present. This process of rejuvenation, according to Cornelia Bryce Pinchot, was what made conservation "a philosophy of dynamic democracy." Her son hoped that the Pinchot Institute for Conservation Studies would become an expression of that same generative impulse.[19]

CHAPTER FOUR

The Inseparable World

We can develop a new conservation ethic, which, if widely accepted in the
coming decades, could provide new and interesting purpose in life. The
purpose is high. It is nothing less than to secure the permanent prosperity
on a green, enduring earth.

—Samuel H. Ordway Jr.

Samuel H. Ordway Jr. was in a philosophical mood. As president of
the Conservation Foundation, and thus a partner with the U.S. Forest
Service in what he and his collaborators considered a "unique cooperative educational venture," Ordway used his speech at the 1963 dedication
ceremonies of the Pinchot Institute to reflect on the present state of the
nation—as a polity and as a land.

He found it wanting because his fellow citizens had not thought hard
enough about their proper place on the planet: "We as people should be
sentient and humble in our way. We know that in the beginning there was
no life on earth; there was no man." Because most Americans only had an
inkling about how "organic life began," what most focused on was how
relatively quickly humans had managed "to dominate the earth." The real
challenge was not to maintain that dominance but to choose to function in
the future in a healthier way: "How successfully man shall now exert our
power, our science, our technology, our wisdom or unwisdom, shall be the
story of our time."[1]

Ordway left little doubt that the choice, and the polarities that framed
our options, were straightforward. Would humans demonstrate the humility necessary to admit that they were only "one organism living in a world
of many interdependent organisms, and that all of these in turn are dependent on air, water, the sun's energy, and the very rocks that form the soil

from which the forests grow"—and from which humanity pulls the "raw materials of sustenance"? Or would "man today, in his self-confidence, bulldoze his way to greater imagined glory—and insecurity, deteriorating as he goes the complex environment on which life forms depend"?

If the latter was their choice, if the citizenry was willing to live in what he called a "synthetic future," putting its trust in technology to resolve all ills, if it was content "to live despite polluted air, polluted streams and seas, on hydroponics and algae and sterilized substitutes for pure air and clean water, on a cement-covered, crowded land," then there would be no reason for anyone to have joined in the celebrations at Grey Towers.

Creating the Pinchot Institute only made sense, Ordway concluded, if Americans generally embraced their integral and integrated position in the larger ecosystem of which they were a part. And the organization would only succeed if its advocates vigorously promoted the collaborative engagement of "citizens, private enterprise, and government" that had brought it into being. The new organization's purpose, after all, was predicated on the belief that "with insight and through education, man and nature and all science and technology together can assure the unity and worth of life on earth."

Although the questions Ordway posed, and the answers he derived from them, may have appeared self-evident, the reality of the task that he and his colleagues took on under the guise of the Pinchot Institute was not nearly as clear-cut as his rhetoric implied. Neither was it as uncontestable as he had allowed. His insights were reflective of a broader debate within American environmental culture about its aspirations and purposes. Ordway was among those calling for a "new conservation ethic" that found its inspiration in anxieties emerging out of the impact a growing population could have on a finite set of natural resources. He brought Malthus to Milford.

The most intellectually pointed of the speeches that fine day in September 1963, Ordway's words depended heavily on the arguments of George Perkins Marsh, whose seminal text *Man and Nature, or, Physical Geography as Modified by Human Action* (1864) had urged its many readers, then experiencing firsthand the blunt, disruptive power of the Industrial Revolution, to become much better stewards of the natural resources that fueled the American economy. If they did not, they would face dire consequences. "We have now felled forest enough everywhere," Marsh wrote, and it was past due that Americans began to "restore this one element of material life to its normal proportions, and devise means for maintaining the permanence of its relations to the fields, the meadows, and the pastures." To do so would require a change of heart, "a certain persistence of character," that once adopted would "help us become, more emphatically, a well-ordered and stable commonwealth, and not less conspicuously, a people of progress."[2]

Consciously or otherwise, Ordway made good use of Marsh's sentiments as well as those who had updated Marsh's assertions about the carrying capacity of the land and limits to (and of) growth, from John Muir, George Bird Grinnell, and Theodore Roosevelt to Gifford Pinchot, Aldo Leopold, and Rachel Carson. As Leopold had pointed out in *Game Management* (1933): "Man thinks of himself as not subject to any density limit," a dangerous fallacy perpetuated by economic and political ideologies, and "industrialism, imperialism, and that whole array of population behaviors associated with the 'bigger and better' ideology are direct ramifications of the Mosaic injunction for the species to go to the limit of its potential, i.e., to go and replenish the earth." Having done just that, Leopold observed, humans were beginning to realize that "slums, war, birth-controls, and depressions may be construed as ecological symptoms that our assumption about human density limits is unwarranted."[3]

Another who shared Leopold's conviction about the perils of prosperity was Fairfield Osborn, a major influence on Ordway's thinking and the founder of the Conservation Foundation. Son of Henry F. Osborn, a prominent paleontologist and president of the American Museum of Natural History, and grandson of William Henry Osborn, a railroad magnate, Fairfield Osborn had retired from business in 1935 to devote himself to the conservation cause. A longtime trustee of the New York Zoological Society, Osborn, along with Ordway and the philanthropist and developer Laurance Rockefeller, among other kindred spirits, formed the Conservation Foundation in 1948; he served as its president until 1962 (at which point Ordway replaced him). The new organization worked effectively behind the scenes and became one of the centers of power within the postwar conservation movement.

It did so in part because of the surge of interest generated by Osborn's 1948 bestseller *Our Plundered Planet*. In it he advanced a neo-Malthusian conviction that the booming postwar American economy and the paired demographic baby boom soon would outstrip the nation's natural-resource base. Osborn worried about Americans' can-do techno-fixation, the untested and unquestioned faith that science would finesse the impact of the nation's runaway consumption; he raised doubts about the willingness of consumers to embrace "artificial substitutes for natural subsistence" and was skeptical that by itself capitalism would mitigate the damage that it too often wreaked on the globe's resources.[4]

Published that same year was William Vogt's *The Road to Survival*. Its critique of the profit motive, which Vogt argued was "divorced from biophysical understanding and social responsibility," was tougher than Osborn's, more sharply critical of capitalism's responsibility for rapid forest devastation, the disappearance of wildlife, and overgrazed grasslands, a chasing after profit that had left behind a "gullied continent." Yet taken together, the two books signaled the emergence of a new phase in the conservation movement. During the Great Depression and the Second World War, the questions confronting conservationists had been framed by intense scarcity and pressing need; through the New Deal and wartime emergency, the notion of governmental planning and careful allocation of resources in general had defined the conservationist agenda. But with the end of hostilities, that time of fear and want was over. The American Century was at hand. Its constituent elements were rapid economic development and a population surge, set within a now-global competition for resources and the militarized pursuit of Cold War hegemony. These factors understandably influenced how conservationists reimagined their central ideals. With the "nation turned increasingly into the Keynesian 'consumer's republic,'" observes historian Thomas Robertson, "Osborn and Vogt began raising flags about the environmental consequences of a rising American standard of living." The ramifications were manifold: through consumption, a new world order was being born, and the two authors cautioned their readers that it "would yield not peace and prosperity, but more war." By connecting "national security with environmental issues," Thomas concludes, "Osborn and Vogt focused attention on their new approach to an old issue—natural resource depletion—and exposed a growing divide among conservationists."[5]

The death of Gifford Pinchot, a major architect of the Progressive Era's conservationism, in 1946 would seem to substantiate this claim; the old order had passed. Except that Pinchot long had been as troubled as Osborn and Vogt would be by industrial capitalism's global reach and pernicious environmental impact. As early as the 1920s, he had written a scathing critique of the Ford Motor Company's massive rubber plantation in Brazil—Fordlandia—for the Pan-American Union (forerunner of the Organization of American States), urging the United States' good neighbors to the south to protect the Amazonian rainforest before it was carved up and cut down as earlier had happened to Uncle Sam's woodlands. After a visit to the Galápagos in 1928, in which he witnessed the impact of an emerging international tourist economy on its imperiled flora and fauna, Pinchot had proposed that Ecuador declare the islands a protected biological preserve. In "Conservation as a Foundation for Permanent Peace," a 1940 article in *Nature,* he anticipated Osborn's and Vogt's postwar musings about the role that the mad scramble for natural resources had played in the two most recent global conflicts, arguing that to stop such plundering would require an international body charged with their protection and management. War may have been "an instrument of national policy for the safeguarding of natural resources or for securing them from other nations," Pinchot noted, but this did not have to be humanity's inevitable fate: "International cooperation in conserving, utilizing, and distributing natural resources to the mutual advantage of all nations might well remove one of the most dangerous of all obstacles to a just and permanent world peace."[6]

In the final chapter of *Breaking New Ground,* Pinchot's posthumous memoir published a year before *Our Plundered Planet* and *The Road to Survival* appeared in bookstores, Pinchot lashed out at the structural changes that were warping the American economy and subverting the people's rights. The intensifying monopoly over resources, he charged, "prevents, limits, or destroys equality of opportunity" and thus is "one of the most effective ways to control and limit human rights." Those who trumpeted the value of "Concentrated Wealth" were no less dangerous. They attributed "the prosperity and progress of the United States" to the free enterprise system, which can only "maintain its strangle hold over the general welfare if it can get the people to accept its exactions, and especially the methods by which it gets its power, as normal and natural." Those who support and expand the free enterprise system at all costs accustom the people to their "tyranny" by discrediting "liberal leaders and movements" through their "ever-increasing control of the press, the radio, and other news outlets." His concluding oration was just as blunt: "The Earth . . . belongs of right to all its people, and not to a minority insignificant in numbers but tremendous in wealth and power. The public good must come first."[7]

With Pinchot's fundamental distrust of the big and powerful, William Vogt would have felt right at home. So to a degree would Samuel Ordway: in a 1961 essay in the *Saturday Review,* provocatively titled "Plunder or Plenty," he explored the intellectual context in which Osborn's and Vogt's books were operating. Among their achievements was the fusion of fears of overpopulation and "widespread undernourishment with the man-and-nature theme of Marsh and the ecological conscience of Leopold." But these authors also, and significantly, aligned themselves with capitalism's discontents, "arousing fear and antagonism in exploiters and tycoons who sensed that their purses, their faith, and their status was being threatened."[8]

Yet neither Fairfield Osborn nor Samuel Ordway accepted the elder Pinchot's certitude that a regulatory nation-state was critical to the defense of the people's rights and nature's bounty from rapacious capital and capitalists. In the emerging age of consensus of the postwar era, their ideological convictions were more cautious and conservative. They and their peers at the Conservation Foundation were convinced that the best way to fix the problems confronting humanity and the planet it inhabited was through dispassionate analysis; expertise could be brought to bear, in combination with corporate self-interest, to better manage the widening gap between production and consumption of resources. The same was true for the giddy, postwar rush of individuals expecting to buy their way into material happiness. Osborn, for instance, hoped that Americans in time would learn "to cooperate with nature, temper [their] demands, and use and conserve natural living resources of the earth in a manner that alone can provide for the continuance of civilization." Ordway added flesh to this bare-bones desire. Conceding that "the optimists are obviously sincere in their belief that technology can feed, house, and turn out unlimited numbers of gadgets for unlimited numbers of men," he and those he dubbed "New Conservationists" came to a different conclusion: if the "demand for more food for more mouths and more machines for more leisure continues to expand at present rates, the limit of expansion will be reached within foreseeable time despite new technological discovery and application."[9]

To forestall that potential trauma, to sidestep "an economic and spiritual reverse of profound significance," would require a new philosophical approach that rested on shifting the modes of consumption. This could not be achieved through government regulation, Osborn and Ordway argued: if "we voluntarily gear our usage of supply of materials to increasing or decreasing supply now, and adjust our philosophy to a possible transition to less material values," Ordway wrote in *Resources and the American Dream* (1953), it would be possible "to win back the security of life which exists where men are working with nature to produce more than they consume."[10]

To achieve this new greatest good required close cooperation between citizens and corporations: "Our needs can be supplied if our wants are bridled, our cupidities curbed. The false ideology which worships unlimited expansion must go. Man must cease to depend for happiness on wealth-consuming gadgets and institutionalized security. Industry itself can limit its own production to the more essential needs of man to keep man free."[11]

Ordway's shimmering vision—freed from the hunger for more, where "we might become happier and more prosperous, extolling not primarily an industrial civilization, nor a return to the primitive, but a balanced civilization on a living earth"—meshed nicely with the accommodating activism of the Conservation Foundation. Unlike the increasingly confrontational tactics of the Sierra Club, as led by its new and forceful executive director David Brower, or the more aggressive lobbying of Howard Zahniser, executive secretary of the Wilderness Society, Fairfield Osborn and Sam Ordway chose instead to remove the Conservation Foundation "from direct advocacy, preferring instead to emphasize through publications and conferences how government and industrial practices could be made more rational. The organization primarily focused on promoting expertise for its educational and policy roles, and thought policy changes were best achieved through the proper application of expert knowledge."[12]

This more behind-the-scenes approach was consistent with the proclivities of those who sat on its board of directors too. Laurence Rockefeller, who embraced Osborn's and Ordway's perception that "education might bring about change without brutal governmental intervention," and who was himself schooled through his work for the foundation about "the importance of scientific data in support of environmental arguments," used his deep pockets to further any number of related conservation causes (and on whose boards many of these same men served). Rockefeller's participation invariably was out of the public eye. The same held true for Roger Hale, who represented the Conservation Foundation on the National Resources Council of America (a clearinghouse for such like-minded groups as the American Forestry Association, Resources for the Future, and the Society of American Foresters), earning this praise from another representative to that body: "He was indeed a gentleman, a man of high ethical standards, not a professional conservationist but one who was ardent in his dedication to conservation." The Conservation Foundation was not a home for rabble-rousers.[13]

Another low-key, dedicated nonprofessional on this tight-knit board was Gifford Bryce Pinchot, who joined it in the early 1960s. Although largely a result of his interaction with the organization in the run up to the creation of the Pinchot Institute for Conservation Studies, he was drawn to its work as well due to a personal affinity with some of Ordway's arguments. In particular his indictment of how gadgetry was debasing Americans' sense of independence and self-sufficiency. As an adolescent Pinchot had become a skilled woodworker, welder, and blacksmith; he also was a champion ocean racer who for more than twenty-five years sailed long distances, in fair weather and foul, without motor, radar, or other "fancy electronics," a preference that meshed with a studied sense of independence and self-reliance: "we were on our own with no artificial aids," he wrote about sailing "blind" through thick fog, unfamiliar straits, or on a transatlantic voyage; "it made our senses sharper."[14]

Like Ordway and Osborn, Gifford Bryce Pinchot also was convinced that the nation needed a more rational approach to resource management, in this case, marine fisheries. As he put it in his speech at the 1963 dedication ceremony of the Pinchot Institute: "There is no need to remind this audience that the population of the world shows every indication of outgrowing its food supply in generations to come." He predicted that an "effective use of the ocean as a food producer will be necessary, not only to protect our parks, forests, and wilderness areas, but ultimately to prevent starvation itself." But as Pinchot cruised the globe's waters, he was struck by a curious anomaly: "In spite of our rapid advance in knowledge about many aspects of the oceans, we are still ignorant on the biological side. This is dramatically illustrated by the fact that in our use of the oceans for food we still live in a civilization based on hunting. We have not yet learned to seed, fertilize, or harvest the oceans as we have the land."[15]

Pinchot's ideas, which he would develop in a series of scientific articles over the next decade, dovetailed nicely with notions that Ordway had sketched out in *A Conservation Handbook* (1949) and *Prosperity beyond Tomorrow* (1955). Writing about fisheries management, its regulatory functions of limiting, when necessary, annual catches and preventing, where possible, effluent and other pollutants from destroying fish populations, Ordway also addressed its focus on "the condition and carrying capacity of given waters." These limitations, like those terrestrial ones, defined (or should define) the capacity of humanity to draw food from rivers and oceans, the first step of which, Pinchot and Ordway reasoned, was to discover exactly what it meant to maintain the marine environment's "ecological stability." Theirs was a meeting of the minds.[16]

These men's reasoning, their shared sense that the intellect, properly trained, could analyze such large questions as food production set within natural systems, was consistent with how the Conservation Foundation pursued its educational objectives. By pushing questions of population and resources to the "center of conservation discourse in the 1950s and 1960s," while simultaneously forging links between "industry, government, and universities in promoting resource management," it set the stage for the new kind of integrative and collaborative analyses that the Pinchot Institute for Conservation Studies was expected to develop and exemplify. By doing so, the foundation also hoped to alter the conversation about how humans could develop what Ordway affirmed was their moral calling: "permanent prosperity on a green, enduring earth."[17]

Under Fire

In this dawning age of abundance more and more Americans will have the chance to experience God's great outdoors.

—Orville L. Freeman

Orville L. Freeman, the secretary of agriculture, flew on Air Force One with President John F. Kennedy, heading north from Andrews Air Force Base to Stewart Air Force Base in New York, from which they would then take a chopper to Milford. Despite the significance of the forthcoming celebration at Grey Towers, the secretary wasn't nervous because of his proximity to the charismatic president. The pair had developed a strong working relationship from the start of the Kennedy administration. A liberal governor of Minnesota before coming to Washington, Freeman had a sharp sense of humor that endeared him to the quick-witted president; when asked why the urbane Kennedy had tapped him for this particular cabinet post, Freeman reportedly quipped: "I'm not sure, but I think it's something to do with the fact that Harvard does not have a school of agriculture." Although Kennedy was "full of enthusiasm" for Freeman, this did not shield the secretary from the chief executive's legendary barbs: "He would frequently kid me about agriculture and always about the budget. He would say that anybody that could spend as much money as I spend in agriculture ought to have a special kind of recognition out of the government. He always had the needle out on this. I think in part it was because he knew it would set under my skin, and he would tease me about it."[1]

At this particular moment, however, Freeman's anxiety had less to do with his relationship with the president—political and personal—and more with his part in the coming dedication of the Pinchot Institute. His intent was to foreground in his speech the central role that the U.S. Forest Service played in the Department of Agriculture and the outsized contribution Gifford Pinchot had made as the founding chief of the public lands agency. "I had some points to make that were important to the Forestry Service that had, in part at least, been left out of this conservation tour," Freeman remembered; "most people in the United States don't know that the Forestry Service is in the Department of Agriculture anyway. This is sometimes a bit of a morale problem. So I had some things to say, since Gifford Pinchot was really chief of the Forestry Service and had done great things in this area." The president was not persuaded: "We were sitting in his cabin in the plane flying up there, and when I handed [my speech] to him he quickly read it and turned to me and said,—'That is fine, but can't you make it shorter?' I didn't make it shorter because there were some things that I had to say. He never said anything, but I think he was a little impatient through it."[2]

True to his word, Freeman praised Pinchot's formative role in establishing the Forest Service, under the mantle of protection and support of President Theodore Roosevelt and Agriculture Secretary James Wilson. Some of what may have made Kennedy squirm was the extensive quotation Freeman pulled from the mission statement that Wilson had sent Pinchot on February 1, 1905, to guide his efforts (a letter that Pinchot had composed for his boss), and according to Freeman, "in classic language it set down the spirit and philosophy that has dominated the Forest Service ever since."[3]

Central to that philosophical perspective was a distinct political realism: "The permanence of the resources of the reserves is therefore indispensable to continued prosperity, and the policy of this Department for their protection and use will invariably be guided by this fact, always bearing in mind that the conservative use of these resources in no way conflicts with their permanent value." That being so, Wilson advised Pinchot (who thus advised himself) that "where conflicting interests must be reconciled[,] the question will always be decided from the standpoint of the greatest good of the greatest number in the long run." Freeman brought this maxim into the present: just before he introduced President Kennedy to the expectant crowd he assured them the president, "the Number 1 conservationist in the United States," was fully in support of the consensual principles guiding the Forest Service's actions for the past sixty years: "President Kennedy's presence here demonstrates better than words his keen interest and firm support for the conservation needs of this Nation. I know first-hand from repeated personal experience that the President is keenly aware that our irreplaceable natural resources must be, at the same time, effectively used to meet the needs of more and more Americans and carefully husbanded for generations to come."[4]

There was only one small tweak to the Wilson-Pinchot memorandum that Freeman acknowledged must be made to make it fully contemporary: "Add to the uses Pinchot spelled out in his multiple use letter of instruction that of recreation, including Wildlife, and this instruction is still operational on September 24, 1963." Freeman's revision is of prime importance, reflecting as it does that these concepts had been encoded in the Multiple-Use Sustained-Yield Act of 1960. This legislative initiative reflected as well that the Forest Service, however unevenly, was adapting to alterations in land-management strategies and public concerns—among them the demand for increased recreational opportunities and a more robust protection of threatened and endangered species. What Secretary Freeman and Forest Service Chief Edward P. Cliff, who served as the master of ceremonies that day at Grey Towers, could not yet fully perceive was just how fast those demands, and a bewildering array of others, would come at them, or how much these pressures would challenge and complicate their work in the coming years.[5]

Emanating from inside and outside of government, through court decisions, grassroots activism, and media investigations, these confrontations and disruptions were the result of a more assertive environmental movement, which took its cues not from Gifford Pinchot and Theodore Roosevelt but from what it perceived to be the more radical insights about the dynamic relations between humans and nature that were courtesy of John Muir, Aldo Leopold, and Rachel Carson. As Freeman and Cliff stood on the platform ready to accept the gift of Grey Towers and open the Pinchot Institute for business, the ground was already shifting beneath their feet.

Yet the Forest Service's history is rife with such turning points, as Gifford Pinchot might have reminded Freeman and Cliff. Implied in his famed tenet—conservation must be focused on "the greatest good for the greatest number in the long run"—was the notion that the definition of the greatest good would, even should, change over time; adaptable too must be the organization that manages the land and its resources by that evolving principle. That is what Cornelia Bryce Pinchot had in mind when she argued at the 1949 dedication ceremonies of the Gifford Pinchot National Forest in

Washington State: "No one understood better than Gifford Pinchot that the battlefields of the future in certain respects must necessarily take different form from those of the past."[6]

The intertwined arc of Pinchot's career and the agency he founded makes that same claim. In 1898, after studying forestry in France and serving as a consulting forester and public advocate for the new profession, Pinchot became the fourth head of the division of forestry in the U.S. Department of Agriculture. Immediately, he started planning for the creation of what would become the Forest Service. He needed first to build public support for the creation of a land-management agency that would regulate the public domain. Hitherto, the public lands west of the Mississippi had been given away, sold cheaply, or lost due to fraud. This privatization was politically acceptable, for the stated policy ambition was that these lands would build western economies and communities and, in time, bring new states into the Union. But the environmental costs of these land transfers, which totaled upward of one billion acres, many of which were prime forested land, intensified in the late nineteenth century. Fearing a timber famine, which could only be countered through federal management and rehabilitation of these valuable if heavily logged lands, scientists and citizen activists started pressuring Congress to act. In 1876 it created the small division of forestry that Pinchot would inherit twenty years later; in 1891 it established the first forest reserves, and between then and 1897 Presidents Benjamin Harrison and Grover Cleveland added nearly forty million acres to the reserves; a rider attached to an 1897 appropriations bill, now considered the agency's Organic Act, defined how those new reserves were to be managed.[7]

To capitalize on these initiatives, Gifford Pinchot and his staff moved in two directions simultaneously: without forests to work on—the reserves were located in the Department of the Interior, while the nation's foresters were in the Department of Agriculture—the agency issued Circular 21, which offered its professional forestry services to landowners large and small. This public-private collaboration gave its agents an opportunity to field-test their ideas and secure favorable press. They also launched a quiet campaign to transfer the national reserves to their care, which received a huge boost in 1901 when Theodore Roosevelt succeeded the assassinated President McKinley. Four years later, the transfer was complete, and the Forest Service was born.[8]

That is when things became complicated. Instituting the ethical dimension of conservation management required a delicate balancing of present and future needs, a novel approach. The first principle, Peter Pinchot has argued, was "to develop the resources for the benefit of the present generation. The second was to protect the integrity of those resources so that you can develop them for the benefit of future generations. And really the most controversial part" was to insure that those "resources were available for the benefit of all the people, not just the privileged few." Enacting this ethos over a broad swath of land increased the pressures. President Roosevelt added upward of 150 million acres to the national forest system during his tenure in office (1901–1909). Forest Service leadership had to develop a multi-tiered bureaucracy, hire employees at all levels, and commence surveying, managing, and providing basic fire protection to these new federal forests. They also developed research stations and nurseries to aid its scientific analysis and regeneration of abused landscapes. To educate the swelling workforce, the Pinchot family donated more than $250,000 to create the Yale Forest School in 1900, and the school operated a summer training camp for its students on the family property in Milford. That same year Pinchot had established the Society of American Foresters and in time the *Journal of Forestry,* vital signifiers of the field's professionalization. As an added mark of these new foresters' expertise, agency staff designed a quasi-military uniform, issued the *Use Book* (1905), describing forest rangers' legal authority to manage the public lands, and developed a code of ethics dubbed "Rules for Public Service" to guide the managers' behavior. To establish legal precedence for its regulatory power, the agency routinely sued violators in federal court (or was sued); each success—and in the early years it never lost—added to its clout. By 1910, the Forest Service served as a harbinger, the novelist Hamlin Garland assured readers of *Cavanagh, Forest Ranger,* of the new nation-state; it was for him a much-needed civilizing force in the rough-and-tumble West, a uniformed protector of the people's forests.[9]

The agency's custodial work continued to inform its land-management strategies for the next thirty years or so, despite the fact that Roosevelt's successor, William H. Taft, fired Pinchot for insubordination in 1910. After Taft had replaced Roosevelt in 1908, the new president and the forester repeatedly clashed because Pinchot did not believe that Taft shared Roosevelt's conservationist principles. News of dubious coal-field leases in Alaska encouraged Pinchot to confront the administration behind closed doors and later in the public arena; Taft had no other course but to set the forester aside. Although the subsequent controversy consumed Taft's administration, crippling his chances for reelection in 1912, and despite the withering blowback—Congress slashed the Forest Service's budget and authorities—the agency continued to operate within the topographical, political, and legal boundaries set at its founding. It maintained as well its central managerial task dating from 1905, which began with regenerating the mountainous West, a scope that widened during the Great Depression to encompass eroded southern farmlands and the windblown prairies of the Great Plains. By the Second World War, the Forest Service's mission was fully national in scope and local in import; it served as America's soft-hatted custodial agent.[10]

When forest rangers started wearing hard hats in the postwar years, their switch in haberdashery signaled a fundamental change in agency mission and action that was symbolic and real. Regeneration of landscapes was subordinated to an accelerated harvesting of timber in many national forests. Less than two billion board feet had been cut on these lands in 1940; twenty years later the number had ballooned to more than nine billion, and, in 1987, at the peak of production, more than twelve billion board feet were felled. This rapid expansion was of incalculable economic importance and also quite controversial, perhaps best reflected in the public uproar in the 1970s over clear-cutting in Montana's Bitterroot National Forest and West Virginia's Monongahela National Forest. For its supporters, clear-cutting signaled the Forest Service's technical skill at harvesting trees on steep-sloped terrain that in the past could not have been harvested; they praised the agency's aggressive effort to transform natural forests into tree plantations. Its critics were numerous and outspoken, and among their ranks was Gifford Bryce Pinchot. For them, clear-cutting was an indictment of the agency's uncritical embrace of the technological fix. They filed lawsuits in federal court, organized demonstrations in various national forests and lumber towns, and pressed for an array of state and congressional inquiries. So fierce did the debate become that it dominated professional journals and popular magazines, leading an internal agency task force to rebuke employees on the Bitterroot for acting as if "resource production goals come first and . . . land management considerations take second place."[11]

Out of this contentious environment emerged a new legal landscape, denoting the third stage in the Forest Service's history. Among its key components was the National Forest Management Act (NFMA) of 1976, which gave the public a much stronger role in determining forest planning and set strict limits on the Forest Service's clear-cutting practices. The NFMA was the last in a remarkable series of landmark environmental initiatives that Congress passed beginning in 1964, within months of President Kennedy's assassination. The Wilderness Act and the Land and Water Conservation Fund—two projects that JFK had promoted—were enacted. Coming online in subsequent years were the Wild and Scenic Rivers Act of 1968, National Environmental Policy Act of 1970, the various Clean Air and Clean Water Act amendments of the 1970s, and the Endangered Species Act of 1973. These bills constitute a pivotal moment in environmental activism in the United States. One ironic, and intentional, consequence of these reforms has been that these new laws regulated the actions of the very land-management agencies, such as the Forest Service and the National Park Service, that had been born during the Progressive Era, an earlier moment of environmental ferment.

Another irony concerned Secretary Orville Freeman's worry that too few of the American people knew anything about the Forest Service, about its sterling history and legendary founder. He need not have been so anxious: within a year of his speech at Grey Towers the federal land-management agency would become a household word (and not in a good way).

This was also the larger context in which the Forest Service Chief Edward P. Cliff had moved up through the ranks. He had joined the agency in 1931 as a range and game management specialist, and by 1939 he had been named supervisor of the Siskiyou National Forest; by the early 1950s he became the assistant chief for National Forest Resource Management. Ten years later, with President Kennedy's and Secretary Freeman's blessing, he was appointed the ninth chief of the agency. For Freeman, the selection seemed natural, later indicating that Cliff seemed the "most chiefly" of the other candidates. This bearing itself may have had something to do with Cliff's mode of leadership: "any chief, to be successful in managing the Forest Service, has to be somewhat of an extension of the traditions and kinds of leadership that preceded him," Cliff asserted, noting that as such no individual "can really shape the destiny of the Forest Service." Cliff did not envision himself as an agent of change, a characteristic that may have appealed to Freeman. If so, this quality also meant that he, and by extension the agriculture secretary, might not be the most nimble navigators through the tumult that would engulf them shortly after their participation in the ceremonies at Grey Towers.[12]

The guiding ideas to which they had pledged allegiance on that stage in

1963 soon were under assault; their flexibility would be sorely tested, and their capacity to respond creatively would be under increased scrutiny. The Forest Service did not fare well. Cliff admitted that he and his colleagues had adopted a defensive posture throughout the sixties and early seventies. In his 1964 annual report, as the first arguments emerged against clear-cutting, Cliff had ignored their implications: "forestry measures must be intensified to meet future demands. Annual timber growth is now exceeding the harvest. But the declining supply of larger, high-grade trees and the overburden of low-grade timber are serious problems. There is no room for complacency as we move to meet the rising demands of all kinds that are being concentrated on the Nation's forest land." The agency also opposed what it considered a precipitous ban on DDT and initially lobbied against the Wilderness Act (as did the Park Service), believing that its constraints limited its managerial options. It also miscalculated the meaning, influence, and growing power of the emerging environmental movement. Certain that only a small number of "militants" were challenging "the whole legal basis for timber management on the national forests," Cliff belatedly recognized that this fight was not simply a "self-centered protest from a very small segment of the population." Rather it represented a watershed moment in American political culture: "What I didn't realize was how potent they could be in expanding this protest."[13]

His successors would have to relearn this lesson time and again. As the legal battles piled up, and the agency lost more often than it won, successive Congresses and presidents proposed drastic budget cuts that led to a diminishing number of Forest Service employees. As the Forest Service faced hitherto unknown population pressures along what it would dub the urban-wild land interface, among them spikes in recreational use, accelerating water-quality issues, and a dangerous uptick in the number of forest fires scorching the parched West, one former chief described the embattled agency as paralyzed.[14]

In this fraught context, the Forest Service may not have been as well equipped as it might have been to nurture the innovative, if untested, public-private partnership it established at Grey Towers with the Conservation Foundation. One of the smaller, though still revelatory dilemmas that the Forest Service would confront in the mid-1960s was what to make (and how to make use) of the Pinchot Institute for Conservation Studies.

CHAPTER SIX

Greening the Presidency

It is a source of interest to me that the three Americans in this century who have been most clearly identified with the maintenance and development of our natural resources and the conservation of those resources, particularly in the West, have been three Easterners—Gifford Pinchot and the two Roosevelts.

—President John F. Kennedy

President Kennedy came to Milford to make a bit of mischief. That was Benjamin Bradlee's later memory of the presidential trip to Grey Towers. Then a *Newsweek* correspondent covering the White House, as well as a presidential confidant—the Kennedys and Bradlees had lived next door to one another in Georgetown before the Massachusetts senator claimed the White House in November 1960—Bradlee's version of events was a tale of three women: his wife Antoinette (Toni) Pinchot Bradlee, a niece of Gifford Pinchot's; her sister, Mary Pinchot Meyer, the former wife of top CIA official Cort Meyer and a onetime mistress of the president's; and their mother Ruth Pickering Pinchot, the second wife of Amos Richard Eno Pinchot, Gifford's brother. It helps to know two additional elements of this story's staging. The Pinchot sisters flew with the president to Milford. His offer of transport was a gesture of respect for those for whom he had warm regard, bringing his friends back to their formative grounds.[1]

As for their mother, Kennedy liked Ruth Pinchot after a fashion though their political opinions could not be more opposed. She and her husband had shared Gifford and Cornelia Pinchot's progressive politics until the early 1930s (Amos, for instance, had been one of the founders of the American Civil Liberties Union, and Ruth had written for the left-wing journal the *Masses*). But with Franklin D. Roosevelt's election in 1932, the couple began to pull hard right. As their criticism of what they believed was the

New Deal's intrusive and paternalistic agenda deepened, Amos and Ruth embraced ever-more conservative positions and organizations. By the end of that decade they were in full support of America First, an isolationist group strongly opposed to the country's entrance into the Second World War.[2]

This rock-ribbed opposition to Democratic Party initiatives continued to define Ruth Pinchot's political calculations when the president, who inherited FDR's foreign policy engagements and faith in social safety nets, visited Grey Towers in 1963. "Her affection for her daughters led her to be more than civil to their friend the president," Bradlee noted, "but it was assumed that every time she saw him she assuaged her guilt by doubling her contributions to Senator Barry Goldwater and to William F. Buckley's *National Review* magazine." Knowing something of this lore, Kennedy reportedly could not resist touring Ruth Pinchot's nearby cottage, where a group photo was snapped. Captured on film, Bradlee recalled, was "one of history's most frozen shots . . . the Democratic president surrounded by the arch-Republican mother and her two Democratic daughters."[3]

This bit of insider hijinks was the real story behind Kennedy's trip to the site, Bradlee inferred, for how else to explain the chief executive's presence at an event that was of relatively little import? How else to make sense of the fact that the president did not bother to tour Grey Towers? About this slight, Bradlee was wrong—Gifford Bryce and Sally Pinchot guided the president through the house and some of its grounds. Yet that truth was less compelling than the journalist's dismissive story line: the gift of the Pinchot family manse "was probably not enough to command the president's presence," Bradlee mused, but "a chance to see where his friends the Pinchot girls had grown up, and especially a chance to see their mother, was apparently irresistible."[4]

As risible, Bradlee thought, was the conservation focus of the presidential tour for which Grey Towers served as the kickoff event: "Except for his love of the sea, John Fitzgerald Kennedy was about the most urban—and urbane—man I have ever met. A well-manicured golf course, perhaps, or an immaculate lawn turned into a touch football field, but that was as far as he could comfortably remove himself from the urban amenities without wondering what the hell he was doing. . . . An outdoorsman he was not." Interior Secretary Udall concurred: "I can hardly, with fairness, complain that my man does not have a streak of Thoreau or Robert Frost in his New England makeup, but I long for a flicker of emotion, a response to the out of doors and overwhelming majesty of the land." More, Kennedy "lacks the conservation-preservation insights of FDR [and] TR, and it will take some work to sharpen his thinking and *interest*." Unlike the interior secretary, however, Bradlee had no expectations that any such heightened appreciation could be nurtured within the heart of the well-polished, loafer-shod Harvard grad. Indeed, the idea that JFK would head off on a "trip across the northern tier of the United States to honor the cause of conservation was an anomaly," he observed, a perspective the White House press corps shared. The moment it learned the purpose of the president's 1963 excursion they coined a series of mocking nicknames for Kennedy—Johnny Appleseed, Smokey the Bear, and Paul Bunyan. "And an unlikelier Paul Bunyan it would be hard to find," Bradlee chortled, "in his well-tailored suits, his custom-made shoes and shirts, walking through the fields and mountains of this land, dedicating dams and parklands."[5]

Juxtapose these contemporary critiques of Kennedy's conservation credentials with a more generous accounting of the thirty-fifth president's environmental reputation. Start with Arthur M. Schlesinger Jr., the court historian whose penetrating prose brought the martyred president so much to life. Accepting that the president was "unregenerately a city man, deeply anxious about the mess and tangle of urban America," he argued nonetheless that JFK held a warm if somewhat abstract appreciation of nature. He fully embraced the intergenerational responsibility "to hand down undiminished to those who come after us, as was handed down by those who went before, the natural wealth and beauty which is ours." The president knew that the fight to insure this transfer was far from easy or simple. Yet his environmental commitments—"Kennedy cared deeply about the loveliness of lakes and woods and mountains and detested the clutter and blight which increasingly defaced the landscape"—coexisted with Cold War tensions, Civil Rights struggles, and the corrosive power of economic injustice. Resolving these hot-button issues held greater priority for the administration.[6]

Schlesinger's measured insights gained additional context in advance of the President's Day celebrations of February 2012, when the journalist Brian Clark Howard published a list of the nation's ten most environment friendly presidents. While he admitted that any such list must be subjective, Howard framed his exercise around whether his subjects advanced the "evolution of environmental policy and protection," with the goal of identifying those "administrations with the best environmental legacies."

Under this criterion, JFK came in tenth: although he used the Antiquities Act relatively sparingly and only on such noncontroversial designations as Buck Island Reef in the Virgin Islands and Alabama's Russell Cave, what caught Brian Howard's attention was Kennedy's convening of a blue-ribbon Science Advisory Committee to investigate the central arguments of Rachel Carson's *Silent Spring* (1962), the best-selling investigation of the impact of DDT and other pesticides on all life forms; the committee largely concurred with Carson's exposé, giving it a presidential imprimatur that helped galvanize public support for increased environmental protections—many of which Lyndon Johnson later signed into law.[7]

So just how green was Kennedy—how verdant was Camelot? Was the man as unreceptive to nature's siren call as Bradlee and Udall suggested? Were his actions as significant as Schlesinger's assessment and Howard's ranking would suggest? Which Kennedy stood up at the podium at Grey Towers in September 1963 to dedicate the Pinchot Institute for Conservation Studies?

The president's speech that day offers some clues about the varied nature of his environmental concerns and about the issues he identified as important to his administration. It reveals as well some of the limitations of that vision, announced as they were at what we now understand to be a critical turning point in American environmental thought and activism.

Kennedy had some sense that he and his audience were living in a transitional moment. "There is no more fitting place to begin a journey of five days across the United States," he declared, "to see what can be done to mobilize the attention of this country so that we in the 1960s can do our task of preparing America for all the generations which are still yet to come." This preparatory labor—to which the newly established Pinchot Institute would turn its efforts—"looks to the future and not the past. And the fact of the matter is that this institution is needed, and similar institutes across the country, more today than ever before in our history, because we are reaching the limits of our fundamental needs of water to drink, of fresh air to breathe, of open space to enjoy, of abundant sources of energy to make life easier." To respond to these pressures confronting urban and rural America would require the creation of new ideas, "the embrace of disciplines unknown in the past." The new organization might bear Gifford Pinchot's name but to fulfill its forward-looking mission, its "active work," of necessity it would draw on a different set of resources and perspectives.[8]

He was optimistic, as presidents tend to be, about the chances of resolving the pressures peculiar to his generation: "I hope that in the years to come that these years in which we live and hold responsibility will also be regarded as years of accomplishment in maintaining and expanding the resources of our country which belong to all our people." A new frontier beckoned.[9]

Yet as the president enumerated the steps his administration had taken or was seeking to take to reach this particular promised land, his resolutions seemed quite indebted to the past. The conservation of water resources meant not reduced consumption or greater efficiency but the ramping up of federal investments in the construction of dams and pipelines to capture and move more water to agricultural, industrial, and urban consumers: among those Kennedy mentioned at Grey Towers were the "Fryingpan-Arkansas and San Juan-Navaho Indian Projects, two of the largest projects of that kind ever approved in a single Congress." Similarly, the growing need for energy required the implementation of new technologies—and here he referred to harnessing steam from the Hanford Atomic Reaction—to generate more kilowatts rather than finding ways to make Americans' energy consumption less wasteful. To meet the booming recreational needs of a society expected to grow exponentially by century's end, this president, as had his predecessors, promised to expand the amount of open space. Kennedy made special reference to an initiative of great local concern and contest, Tocks Island Dam, which would impound the Delaware River just downstream from Milford, and in these waters its residents, among so many others, would be able to swim, fish, boat, and play. Although the president believed that the linking of science and technology to conservation would be a decided benefit for the American people, and marked a distinct difference from the beliefs and actions of earlier generations, in fact he was following the trail that Pinchot and Presidents Theodore and Franklin Roosevelt had blazed.[10]

The same indebtedness appears in the other speeches President Kennedy delivered on his post–Grey Towers conservation tour. The trip was the result of many months of discussion inside the White House between relevant cabinet secretaries and key supporters in Congress. One of these latter had been Senator Gaylord Nelson, who in 1970 was the architect of the first Earth Day. As part of his 1962 campaign for the U.S. Senate, Nelson had conducted his own conservation tour of Wisconsin and had been stunned by the large and enthusiastic crowds it generated; he predicted to Schlesinger that the same would occur if the president embarked on a similar excursion. In a subsequent memo to Kennedy, he wrote: "The fact that you are going on a nation-wide tour will command great attention for several reasons, including the fact that no President has done exactly this before. The question is how to maximize the effect—how to hit the issue hard enough to leave a permanent impression after the headlines have faded away—how to shake people, organizations and legislators hard enough to gain strong support for a comprehensive national, state and local long range plan for our resources." Out of these deliberations it was decided that the president would embark on a major swing across the northern tier of the nation, with a special focus on conservation and natural resources. His itinerary would be crushing: the five-day, eleven-state, fifteen-speech swing would take him from Pennsylvania to Wisconsin, Minnesota, North Dakota, Wyoming, and Montana; Washington and Oregon; California, Utah, and Nevada. In addition to Grey Towers, his scheduled stopovers included such remarkable landscapes as the Apostle Islands of northern Wisconsin (not then a National Lakeshore) and Lassen Volcanic National Park in northern California; he would visit the massive hydroelectric dams of Faming Gorge and Grand Coulee, along with the Hanford nuclear facility on the Columbia River; and his tour would conclude with speeches in places as impossibly disparate as the spiritual center of the Latter Day Saints, the Mormon Tabernacle in Salt Lake City's Temple Square, and that mob-controlled, and Rat Pack gambling Mecca, Las Vegas.[11]

Everywhere the crowds were energetic, even in Republican strongholds. The president had joked about a similar trip one year earlier, that it was a decidedly nonpartisan excursion: "I'm not going to a single state I carried in 1960." The 1963 tour had a more intentional political purpose—to rally support for Democratic representatives and senators facing tight elections the next year. The current occupant of the White House would see some electoral gain as well, argued Senator Nelson: "The fact that you are going on a nation-wide tour will command great attention for several reasons," he wrote Kennedy in late August 1963. While on the road, the chief executive reached a similar conclusion: he and his aides believed that the sizeable crowds and extensive press coverage were marks of widespread support that should give him an advantage in the 1964 reelection campaign. The land just might benefit too: so that those "who come after us will find a green and rich country," the president noted in a post-trip letter to Rep. Wayne Aspinall, the all-powerful chair of the House Interior Committee, he vowed to "mount a new campaign to protect our natural environment."[12]

Kennedy made similar claims in his many speeches during his whirlwind tour, each a rephrasing of and expansion on the themes first developed at Grey Towers. The conservation of soils and water was essential to a productive America; reclamation projects would and should power the growing West. Nature should be controlled, parks set aside, air and water pollution regulated. Only Kennedy's final speech in Las Vegas strayed from this well-worn script, and it did so because Secretary Udall, who had been the president's fellow traveler for the entire tour, rewrote portions of the text to give it a more ecological tone. Declaring himself to be in favor of a "a third wave of conservation in the United States following that of Theodore and Franklin Roosevelt," President Kennedy promoted the passage of a land and water conservation fund then bottled up in Congress, funds from which the government would purchase wetlands, wilderness, and other threatened terrain. "There isn't very much that you can do today that will materially alter your life in the next three or four years in the field of conservation," Kennedy declared, "but you can build for the future. You can build for the seventies, as those who went ahead of us built for us in this great dam and lake that I flew over today. Our task . . . is to make science the servant of conservation, and to devise new programs of land stewardship that will enable us to preserve this green environment."[13]

What is striking about Kennedy's stress on scientific management and on the technological fix is that these emphases, which had held such sway since the late nineteenth century, were already coming under greater critical scrutiny. Rachel Carson was not the first to point out that the scientific enterprise could manufacture as many problems as it seemed to solve. Conservationists were increasingly leery of an engineering ethos that led to the damming of rivers to provide hydropower and irrigation downstream, but that did not account for the environmental costs associated with these massive reclamation projects. In the coming years others would doubt that nuclear energy was quite the liberating force that Kennedy made it out to be. This modern environmental movement, which had emerged in the 1950s, began challenging in the courts, legislatures, and civic arenas the kind of initiatives that the president promoted in his speech at Grey Towers. Among those projects that ultimately did not get constructed due to the public's pushback was the Tocks Island Dam on the Delaware River, for which President Kennedy had had high hopes, knowing it would have created "the largest federal recreational area in the East." In the end it failed to pass muster, as more and more Americans agreed with Rachel Carson's admonition, conveyed in a letter to Interior Secretary Udall, that humans must remember that they are "custodians not owners of the earth."[14]

Her blunt reminder is instructive, illustrating the cautious approach that characterized the conservation agenda of the president and the interior secretary. Their moderate perspective also may explain why, for all the immediate publicity and goodwill the conservation tour generated, its long-term value is ambiguous. To the great disappointment of Wisconsin Senator Gaylord Nelson, the president did not announce the designation of any new parks or landmarks (though the Apostle Islands and the Oregon Dunes were then being bruited for incorporation into the National Park system). He did not assert his unqualified endorsement of the Wilderness Act, still a subject of intense congressional debate, though he had made his support known to the Congress. Instead, the president, like his attentive audiences, was much more captivated by the news of the successful completion of the Limited Test Ban Treaty that broke while he was on the road. "The trek of Paul Bunyan through America," journalist Bradlee concluded "never was much of a story."[15]

Bradlee's cutting conclusion is too sharp. Although JFK was no Teddy Roosevelt, and his administration never embraced the "earth-oriented perspective" that has been so fundamental to modern environmentalism, arguing instead that "resource development, especially dam-building, was essential to economic growth, national security, and political success," Kennedy's record was substantive. As to why this is true, look no further than to the president's very urbanity that had led Bradlee to discount Kennedy's conservation commitments or to what Schelsinger described as his considerable anxiety about "the mess and tangle of urban America." The nation's teeming cities of the early 1960s, their outward sprawl and intensifying concentration, dominated Kennedy's environmental agenda. A case in point was his designation of the nation's first three national seashores on Massachusetts' Cape Cod, Padre Island in South Texas, and Point Reyes, just north of San Francisco. These sites not only were selected to represent the three major bodies of water that shaped the country's continental coastline (the Atlantic, Gulf of Mexico, and the Pacific), but they were also set aside because they lay within several hours' drive of burgeoning metropolises (the Northeast megalopolis, Houston and San Antonio, and the Bay Area). Their establishment, the president had argued at Grey Towers, was thus a matter of social equity: "I do not know why it should be that six or seven percent only of the Atlantic coast should be in the public sphere and the rest owned by private citizens and denied to millions of our fellow citizens."[16]

The United States had to get ahead of the urbanizing implications that the baby boom posed—"We are going to have 300 million people by the end of this century," Kennedy warned at the 1962 White House Conference on Conservation, the first such confab called in more than fifty years, "and we have to begin to make provisions for them. We do not want, for example, this eastern coast to be one gigantic metropolitan area stretching from north of Boston to Jacksonville, Florida, without adequate resources for our people." To secure some of those resources, Kennedy's administration pushed for congressional funding through the Housing Act of 1961 for advanced water-quality projects, sewage-treatment facilities, and urban parklands. Coordinating these and other proactive measures was the Bureau of Outdoor Recreation, the creation of which Kennedy called for in a March 1962 special message to Congress. It began its work the next year, joining with the Pinchot Institute, to offer remedies through its policy analyses and conservation education programming, and to build a consensus for social change. "Government must provide a national policy framework for this new conservation emphasis," Kennedy had asserted in Milford, "but in the final analysis it must be done by the people themselves."[17]

Interior Secretary Udall surely exaggerated when he opined that these many initiatives collectively might "*easily be the high-water mark for conservation in the history of this country.*" Yet his enthusiasm, especially when read in the context of his doubts about President Kennedy's lack of empathy for the natural world, is not without merit. Kennedy intuited, more than midwesterners Truman and Eisenhower, and perhaps more than Arizona-born Udall himself, that the American present and future were inescapably urban. Its identification of the pressing needs of mushrooming cities, as physical environments and population centers, was the Kennedy administration's real contribution, and it is what set it apart from its immediate predecessors. With the "exception of Lyndon Johnson's Great Society," argues historian Thomas G. Smith, Kennedy's tragically shortened first term was still "the most conservation-minded administration in the postwar era."[18]

INSTITUTIONAL CHANGE

CHAPTER SEVEN

Conservation Education

If this institute does not succeed in training the American people to use its resources wisely, this country will collapse.

—**Maurice K. Goddard**

Dr. Goddard is a prophet. I completely agree with him.

—**Matthew J. Brennan**

The four-fold blessings bestowed on Grey Towers by the Pinchot family, the Conservation Foundation, the Forest Service, and the Executive Office of the President were complicated by the fact that each institution sought to shape the celebratory moment and contribute to a projected future in which the nation benefitted from the Pinchot Institute's success. Yet the impact of the speakers and the entities they represented was also limited by their past beliefs and contemporary enthusiasms, constraining just how much the institute could achieve.

None of these limitations bothered President Kennedy's speechwriters (nor should they have). After all, the chief executive was in Milford to accept the Pinchot family's gift, to give it his benediction, and to lift up the crowd and their sense of Grey Tower's prospects. His closing words did exactly that.

As his keynote address at Grey Towers reached its climax, the rhetoric intensified. Bringing to the fore one of Gifford Pinchot's most forceful, historically evocative arguments about the ineluctable link between environmental protections and social uplift—"a Nation deprived of liberty may win it; a Nation divided may reunite; but a Nation whose national resources are destroyed must inevitably pay the penalty of poverty, degradation, and decay"—Kennedy agreed with Pinchot that conservation could save the day. "Conservation is the *key* to the future, and I believe our future can be bright," the president affirmed. But the level of the nation's success depended on the citizenry's shouldering its responsibilities, on the fiscal commitments and political will that local, state, and federal governments must muster to reclaim a battered terrain and the people who depend on its bounty, and on whether or not the Pinchot Institute for Conservation Studies and its like could generate the requisite "insight and foresight" necessary to advance the frontiers of knowledge. If his audience joined him on "his journey to save America's natural heritage," a "high purpose unparalleled in the history of the world," then, Kennedy predicted, they would assure "a fuller, richer life, for all Americans now and for generations to come."[1]

With cheers raining down, the platform party rose to flank the president who, with Gifford Bryce Pinchot at his side, leaned over the bunting-draped front railing and pulled a thick cord to unveil a large boulder with an embedded brass plaque commemorating that day's celebration and the institute's new motto: "For greater knowledge of the land and its uses." At that, Kennedy flashed his patented smile, waved his hand, walked up the hill, and disappeared from view.

His electric presence lingered. "For many Milford residents the visit of the president had a story book flavor," a journalist wrote, "but the familiar figure of the TV screen and magazine cover was real. He was here in their midst addressing them, talking about their neighbors, the Pinchot family, emphasizing this state, this town, this house." That house, this place, bore a new responsibility, wired in as it was now with the main currents of American life: "Much is left to be done, the President reminded his listeners. The Conservation Institute at Milford, at Grey Towers, the home of Gifford Pinchot will lead in the effort to see that it is done." Elated, "a spiritual buoyancy . . . was lifting Milford into a niche from which it must not fall."[2]

The elation could not last. The Pinchot family felt it first, and one of the most aware of the deflation was also the person most physically close to the scene: the sharp-eyed Ruth Pinchot. By political allegiance and personal temperament, she distrusted the government's ability to safeguard Grey Towers, the site and its treasured artifacts. The Forest Service had assured Gifford Bryce that the family could take whatever mementos they wished, but the Conservation Foundation and the federal agency, expecting to develop a museum devoted to Gifford Pinchot and the larger conservation movement, preferred that they would leave the heirlooms; the Pinchots complied. From her front porch, Ruth Pinchot watched in dismay as many of the family's invaluable collections—Gifford Pinchot's priceless jades, a

George Washington–signed mirror that his grandmother had owned, Hudson River School paintings, which were James Pinchot's priceless patronage, and a host of other furnishings and mementoes gathered over five generations—were packed up so that construction crews could begin to take apart the house. They bulldozed one unsafe outbuilding to make way for a new parking lot behind Grey Towers, shored up termite-ridden beams, and ripped out most of Cornelia Pinchot's gardens and other plantings in favor of low-cost, low-maintenance landscaping, a utilitarian clean sweep. Yet when the house reopened in 1964, many of its most cherished objects never made it back. The artwork was sold at fire-sale prices, and other pieces had vanished without a trace. Ruth reported to her nephew that she witnessed fully packed private cars leaving Grey Towers, a pillaging that the family reported to the Forest Service to no avail. The memory of this crude treatment of the site and its treasures long rankled the Pinchots, as the "uniqueness of the house," Gifford Bryce Pinchot recalled in 1986, "was largely destroyed."[3]

The day-to-day workings of the institute, touted as the new center of knowledge dedicated to producing a robust education curriculum for children and adults, also seemed to have gotten off to a shaky start. Or so Ruth Pinchot surmised. "Whatever happened to the Pinchot Institute for Conservation Studies?" she asked its on-site director Matthew Brennan after returning to Milford in April 1964; "all that shows is a new asphalt road, a tank full of undrinkable water, and an empty [nineteenth-]century house. O, yes, some fruit trees pruned."[4]

Her puzzlement is understandable. At the start of that year, the Forest Service and the Conservation Foundation were still tinkering with a formal agreement about how they would manage the site. "We had thought of the Board of Governors as the top level policy group," Forest Service Chief Edward Cliff wrote to Samuel H. Ordway Jr. in late January, "with the Executive Committee responsible for the actual staff job of implementing the recommendations of the board." Appreciative of Ordway's desire to expand the board of governor's size so as to engage "a few more outstanding and influential names . . . from the standpoint of obtaining financial aid," Cliff wondered whether more bodies meant more money and was convinced in any event that a top-heavy administration "may present problems in efficiency of operations." It took until June to resolve these organizational issues.[5]

Another five months passed before the board of governors held its first meeting, at the University Club in New York City with Chief Cliff serving as its permanent chair. Joining him around the table was a goodly number of conservation policy heavyweights. Among them was Laurance Rockefeller, then chairing the U.S. Outdoor Recreation Resource Commission; Gifford Bryce Pinchot; Samuel Ordway; David Heyman, head of the New York Foundation; Samuel T. Dana, dean emeritus of the University of Michigan's School of Natural Resources; Joseph Fisher, president of Resources for the Future, a policy shop based in Washington, D.C.; and Fairfield Osborn. These men—and it was an all-male board—"reaffirmed the widespread need for conservation teaching and for an acceptable conservation ethic," the development of which was essential if the country hoped to manage the staggering number of "scourges," including polluted air and water, urban blight, declining water quality, and mismanaged lands, that now "affect our mode and manner of life."[6]

Leading the work at Grey Towers were the institute's codirectors, Mathew J. Brennan and Paul F. Brandwein. Brennan came late to the field of conservation education. With a PhD in biology, he had a strong interest in science education, and in 1957 he had identified what for him was the formative notion that must underlie conservation education programming: "Conservation, to me, is a way of life, a philosophy of living based on the natural and physical laws of science and tempered by the moral, intellectual, and social environment of the individual." If it is a mode of living, then it must be inculcated early; earlier, even, than elementary school, but as soon as a child is "old enough to see and recognize beauty." Finding a professional outlet for his insights proved difficult. During the late fifties, Brennan served on a series of polar expeditions to Antarctica, was an adjunct biology professor at Columbia and Rutgers (where he may have come into contact with his counterpart, Paul Brandwein), and for a time worked for the U.S. Office of Education. These varied work experiences shaped Brennan's belief that scientists and conservationists must be in dialogue with one another. "Conservation is science. It is based on all of science," but unlike science as a discipline it had not forgotten its social obligations: "Society owes a great debt to science. But science owes a great debt to society—a debt on which payment is long overdue." One way scientists could embrace their responsibilities to the commonweal was by collaborating with conservationists in their effort to protect and steward the land, and the possibility that at Grey Towers he might facilitate this mutual engagement would be one of his new job's great attractions.

Not that he knew anything about this prospect when he joined the Forest Service in 1961 and was assigned as a junior staffer to the agency's fledgling Conservation Education program in the Washington Office; he would be promoted to its director early in 1963. Later that year Chief Cliff would tap him to manage the agency's activity at Grey Towers, despite the fact Brennan had little administrative experience. This lack, when combined with his sort tenure in the agency, meant he had little working knowledge about its operations and not a lot of personal resources on which to draw to launch this high-profile project. Troubling too was his self-aggrandizing touch—in a brief memoir written in the wake of President Kennedy's assassination, Brennan claimed he had suggested that the chief executive kick off his conservation speaking tour at Grey Towers; he was "thrilled" at the opening ceremonies when the president "spoke some of the words I had written." After he left the Forest Service in 1968, Brennan promoted himself as the institute's sole "founder," a self-regard that would complicate the Pinchot Institute's fitful first five years.[7]

His colleague Paul Brandwein was no stranger to high self-esteem, but, like Brennan, he never had managed an operation of the size and scope projected for Grey Towers. He too had spent many years as a biology teacher and public lecturer and for more than a decade had served as educational director at the Conservation Foundation, where he had pulled together conferences and solicited or wrote articles and chapters on the need for an enhanced conservation education program in America; these endeavors were quite similar to the work that he proposed to continue through the institute. He had a certain following too. "In the 1950s, when conservation became fashionable in polite society, Brandwein had been lionized," observed Calvin Stillman, who had worked with Brandwein at that time through his own work developing environmental studies at Rutgers University. "A somewhat austere individual, he moved easily among the movers and shakers," a reflection of his position at the equally well-connected Conservation Foundation; until the mid-1960s, the glad-handing Brandwein was thought of as "Mr. Conservation."[8]

That might be giving Brandwein too much credit, for surely David Brower, executive director of the Sierra Club, Howard Zahniser at the Wilderness Society, or many of the other influential figures associated with the contemporary Natural Resources Council of America had wider networks, greater achievements, and much larger reputations in the postwar conservation movement. That is not to suggest that Brennan's and Brandwein's work was insignificant—it was important. Indeed, as codirectors at Grey Towers they had been given a new and potentially major stage on which to enact serious pedagogical initiatives, an important space from which to reconceive how students of all ages learned about the environment and how they would apply that knowledge to the better management and more efficient consumption of natural resources. In Brennan's telling President Kennedy had spoken to him directly about the value of their shared aspirations for the institute, assuring him at the Grey Towers celebrations in 1963 that "your work here could do so much for the Nation's problems of idle lands and idle youth."

That charge came with a series of difficulties, some that Brennan and Brandwein could anticipate and a number they could not. Still, the two educators were eager to get started. Brandwein assured local media in September 1963 that the institute had a "three-fold purpose": "We are here to teach, do research, and disseminate information." Brennan agreed: "We are not interested in what I was doing in 1958," he told an inquisitive reporter seeking his biographical background, "the important thing is what we will be doing at this center next year, and the next decade."[9]

By that standard, the institute was a failure. At the start it appeared to have the appropriate resources—human, fiscal, and imaginative—to make its mark. Yet as its early planning documents reveal, the ambitions of its directors and of their sponsoring institutions complicated their productive capacity. In a spring 1964 letter nudging the Forest Service to complete what he considered to be the disturbingly slow rehabilitation of Grey Towers, Samuel Ordway pointed out the obvious: until the site was prepared to hold conferences, little of its real work could occur. Waiting in the wings was an international gathering on wildlife. "Our concern must be the fate of our wildlife in the year 2000," Brennan argued in a prospectus for the proposed event. "Where will our plants and animals go?" Brennan asked. "If we don't start doing something about it now, it will be too late at the end of the century." Another workshop was scheduled to bring together a small coterie of "elder statesmen of conservation," whose task would be to speculate about the movement's work over the next two decades. Brennan and Brandwein announced as well that they intended to hold annual summer seminars for elementary and high school teachers specializing in science education. All of these events, along with the opening of Grey Towers for one-off conferences that other groups would sponsor there, were expected to fill the institute's first calendar year. Its directors dreamed big.[10]

One of the projects that lived up to their outsized aspirations was the January 1966 planning meeting of youth representatives to that summer's National Youth Conference for Natural Beauty and Conservation. The brainchild for the larger event was Diana MacArthur, niece of Lady Bird Johnson; the First Lady was a powerful advocate for the beautification of the nation's highway system, its urban parks, greenways, and scenic wonders. "I never cease to feel that I am part of a new frontier here," Johnson wrote in her diary that summer; "there is so much happening, so much important work of infinite promise whose results I hope to see someday." Johnson envisioned "using the White House as a podium—hopefully—to thank, applaud, to advertise, to rally citizens to action in improving our environment," and MacArthur's project gave her just the chance to take full advantage of the unparalleled platform that her husband's presidency gave her. The plan was to bring five hundred teenagers associated with such organizations as the Girl Scouts, Future Farmers of America, and the Boy Scouts to the nation's capital for a four-day symposium on the need to make America the Beautiful, beautiful again.[11]

To further this group's work, the Pinchot Institute organized a training session in advance for twenty youth leaders who had been invited to attend the national conference. "You have been given a rather awesome responsibility," Brennan declared in his opening remarks on a snowy, late January evening, "representing, as you do, the 20 million members of America's ten leading youth organizations." Thus the Pinchot Institute staff and outside facilitators were there to guide the teenagers in developing a set of topics and goals so that after the June conference they could implement them on a state-by-state, town-by-town basis. While conceding that if "the future belongs to you, then we cannot plan it for you," Brennan and his colleagues already had mapped out a series of programmatic ideas for the conferees to consider as part of their action plan. As it turns out, these were derived largely from the preceding adult conference; the adults only appeared to step aside.[12]

Most of the adult-approved initiatives were feel-good in orientation and import: litter control, roadside beautification, conservation education, and the protection of open space. They also embodied a vision of America's youth working within the system, not challenging its fundamental assumptions about industrial production, global distribution, and mass consumption. None contained the essential critiques of modern capital and power then animating the activism of other young people engaged with environmental issues through the Sierra Club, for instance, or more radically through Students for a Democratic Society. Then again, this was going to be a White House conference, not a mass protest streaming down Pennsylvania Avenue.

Like the First Lady, who in June welcomed the delegates to the South Lawn with words of encouragement—"you will not have reached maturity until you have tackled a hopeless, idealistic cause . . . [which] you may be surprised to find . . . was not so hopeless after all"—Matthew Brennan set great expectations on such small steps. Speaking of the problem of litter, he wondered why "can't the 20 million youth in your ten organizations start a quiet, effective campaign of good example to focus our nation's attention on what they are doing to themselves and their country?" Rather than scold the American people to pick up their trash, "show them! It will take a pretty callous person to ignore the message of the scout or club member who quietly picks up his litter, or steps up and says, 'Mister, I think you dropped this.'" By this simple means, he assured his audience that "you can help them develop an ethic of conservation and natural beauty."[13]

Giddy though he was about the role of the Pinchot Institute in the planning session—"We . . . are proud to have been able to give this great effort the push which was necessary before there could be movement"—Brennan noted what had become a disconcerting pattern: "Mr. Brandwein was in California and Mr. [Cliff] Emanuelson [their assistant] on a New England trip during the period prior to the conference, I so handled the planning and logistics for the meeting alone." For the whole of that year, Brandwein would only be at Grey Towers a total of six days, so it is little wonder that Brennan was feeling bereft. His codirector's absences were of a piece with his seeming indifference to the institute's self-proclaimed goals. The summer teacher conferences that Brandwein insisted were critical to its larger mission were never held; he repeatedly canceled them, citing his dissatisfaction with the number of teachers interested in attending and in the quality of the facilitators who would run the program. From Brennan's perspective, this string of cancellations cut to the quick because these workshops were to have been "the heart of our effort to put into curricula in state and local school systems the fruits of our policy and practice conferences." Their loss was bad enough, but the "policy and practice" gatherings that were supposed to be the foundation of the institute's pedagogical research and development did not occur either, or at least not with the consistency that Brennan and Brandwein had promised. Symposia on wildlife and on "pollution, poverty, population, and pesticides"—the apocalyptic quartet of mid-1960s environmentalism—were cancelled, a track record that alarmed the Forest Service.[14]

Those events that were held might have concerned its leadership for another reason. Brandwein was a more ethereal thinker than Brennan; he loved to engage in big-picture hypothesizing. Bearing this out in part was his pre-institute work on the nature of creativity and the social and psychological dimensions of children's lives that might transform gifted kids into distinguished scientists. So also with the question that seemed most to animate his efforts while with the institute: What was the most relevant form of education "in this moment in history when comprehension of the fittedness of the environment and the interdependence of environments, and of man is essential to securing sanative environments and sane men?" The concept of a "sanative environment" left Gifford Bryce Pinchot scratching his head, wondering why it "seems so popular in [Brandwein's] and Matt Brennan's writings." Pinchot asked some of his colleagues at Johns Hopkins University what the term meant to them and they were "completely stumped. To most laymen it seems to mean something to do with sanitation. The dictionary defines sanative as curative. Can they mean they want a curative environment?" Some younger educators, however, were drawn to Brandwein's insights and their elliptical form of delivery. Charles E. Roth, director of education at the Massachusetts Audubon Society, later noted that Brandwein's key realization was that "our settlement patterns had changed over the years away from rural and small town living to urban living. He knew that urban settings were often unhealthy and that we had to put people into conservation of the entire environment, not just of natural resources." Because Brandwein appeared to have a "much broader view of what should be included under the rubric of 'conservation' than did many of the other participants" at institute-sponsored events, the discussions he led at Grey Towers were "often mind-blowing."[15]

They were also circular. Each one struck the same philosophical pose— that the need to reach children and adults, in and out of schools, was of paramount importance, and was so because "the environment would not be conserved unless people were conserved first," language that Brennan credited to Gifford Pinchot. He "used to say that his chief job as Chief Forester was educating the people. Every conservationist since Pinchot felt the need for an informed public. To date we have not achieved this." As such, each Grey Towers workshop concluded that the assembled educators knew too little about how people experienced the environment, urban and rural, or how they made sense of those experiences. That being the case, trying to figure out how better to educate them about their responsibilities to those environments would be time consuming. Each event then recommended that much more research must be conducted about what kind of conservation education was being taught in the schools, a necessary first step before tinkering with or dreaming up relevant pedagogy: "It would be easy for us at the Institute to sit here and say, 'This is what we are going to do.' But we could be dead wrong. We could go too fast. We could go too slowly. We could go in the wrong direction." Those caveats demanded caution: "What we do here will affect fifty-four million people in the schools each year," Brennan and Brandwein wrote, reflecting on the import of an October 1966 conference on the Techniques of Teaching Conservation, and as such "we must take time to think this through."[16] They did not have that luxury, however. The Forest Service was disheartened by what it sensed was a flawed process, however intellectually merited—each conference seemed to require another to reach the same conclusion. Brennan's episodic reports to his supervisors must have given further pause: there appeared to be little follow-through on the stated recommendations, little action for all that talk. The agency began to doubt the project's viability.[17]

The Forest Service also began to question the Conservation Foundation's commitment to the institute, and with reason, for the foundation, long accustomed to the clubby atmosphere of its New York–based engagements, was undergoing a revolutionary transformation that would drive a wedge in its relations with the federal agency. In 1964, Samuel Ordway stepped down as its president and Fairfield Osborn began to court Russell E. Train as his replacement. Working as a federal tax judge in Washington, D.C., Train, the future administrator of the Environmental Protection Agency, had a long record of achievement, first demonstrating his organizational abilities after establishing the African Wildlife Leadership Foundation, a rapidly growing group that Osborn earlier had tried unsuccessfully to merge into the Conservation Foundation. On the board of several major environmental organizations, Train's fund-raising prowess was well known to the close-knit community that he and Osborn moved within. Approaching Train to take on Ordway's position seemed almost preordained, the biographer J. Brooks Flippen argues; "it all made sense."[18]

Except that Train made some demands Osborn did not immediately accept, most relevant was that he would not relocate from Washington,

D.C. If Train were to become the foundation's CEO, its headquarters must move south. Not wanting to uproot was only part of Train's calculations. He recognized that the nation's capital was the growing seat of power in environmental matters, evidence for which was daily manifest in the slew of conservation legislation then emanating from Congress in the tragic aftermath of President Kennedy's murder. He may not have known it but his calculation had a legacy. In 1909 Gifford Pinchot founded the National Conservation Association to track, influence, and challenge natural resource legislation, and he determined that the only place from which to complete this work was in Washington; in the late 1930s the founders of the Wilderness Society had reached the same conclusion. Fairfield Osborn must have understood this logic, for he accepted Train's conditions, and in September 1965, the Conservation Foundation opened its office on Connecticut Avenue.[19]

Its presence there might have facilitated a closer relationship with the Forest Service and the Pinchot Institute, but it did exactly the reverse. Once considered an accessible asset as long as the foundation was located in New York City, Grey Towers was now a distant liability. That also may have been true of Paul Brandwein, who lived in New York State, about twenty miles east of Milford. In a letter to Gifford Bryce Pinchot, and later in a statement to the board of governors, Train expressed his impatience with Brandwein's grandiose, expensive plans and his conviction that it would take a decade or longer before any measurable impact could be discerned. As Train began to put his stamp on the foundation, Brandwein took the hint, resigning as codirector of the Pinchot Institute and his educational directorship at the foundation.[20]

Brandwein's departure was tied to two other issues that boded ill for the institute: Brennan reported the foundation had encumbered but $5,000 for institute-based projects in 1967, a paltry sum that led him to scuttle all planned conferences. The other troubling news was that the Pinchot Institute's board of governors had stopped meeting. Brennan was not alone in suspecting that this latest turn of events was a deliberate attempt to undercut the work at Grey Towers. In February 1967 he informed the local media that the Conservation Foundation's plans to establish its own task force to study conservation education's future course set up an apparent conflict of interest. In private he was more blunt: "Not only is the Conservation Foundation not going to support a program of the Pinchot Institute (and did not even when Paul [Brandwein] was here)," he confided to Wilson Clark, a member of the institute's board of governors, "but now they are proceeding to develop our task force, and, I understand, also our Conference on Conservation and the College Program. . . . We thought we were only suffering from a lack of support, but now it appears we are also in competition. The failure to include me on the task force makes this fairly obvious."[21]

Russell Train would only concede in a May 1967 letter to the board that "the ambition of the Forest Service and the Foundation for the Institute as a national center for the improvement of curriculum has not been achieved." This, he noted, would be the sole subject of the board's June agenda, a focus that he and Chief Edward Cliff had agreed on in a "candid discussion," which had led to the formation of a four-person subcommittee to determine whether and in what fashion the institute "can best serve the general objective of conservation." This announcement, Gifford Bryce Pinchot believed, was at odds with the board's advisory function as outlined in the institute's charter. If its purpose was to advise, then why had Train studiously ignored several letters he and other members had sent him about the institute's faltering operations? The board's powers were usurped as well by the ad hoc committee that had been formed to devise the institute's future course. "If you decide that the board of governors is to have a function other than that of being figure heads," Pinchot advised Train, "I'll be happy to ride out my term and see what happens." He was not willing, however, "to be in a position of responsibility with respect to the Institute without having the opportunity to carry out this obligation."[22]

He would not remain in limbo for long. The June 1967 board meeting, held in Washington, D.C., provided an explosive climax to the short history of the Pinchot Institute. Forest Service Chief Cliff laid out the agency's lengthy list of concerns, starting with the striking imbalance of support for the institute. The federal agency had invested an estimated $468,000 on rehabbing the Grey Towers and its grounds and expended another $270,000 on salaries, supplies, and travel, for a total commitment of $738,000. By contrast, the Conservation Foundation had put in roughly $160,000 for salaries and programming, an amount that had dwindled each year since its original outlay in 1962; one board member later calculated that the foundation had expended no more than 4 percent of its investment on programming, which, according to the 1964 memorandum of understanding, was its responsibility alone. These stark differences in their fiscal contributions would mean little, Cliff counseled, if the two organizations reaffirmed their shared faith in the Pinchot Institute's purpose: "We esteem the name and the man and all he stood for and accomplished. This institution that bears his name; this historic setting for so much of his work and life; this rich heritage of the ancestral Pinchot home as a central facility—all of these can combine to create a unique opportunity that could not be duplicated elsewhere."[23]

Cliff's oration did not persuade Train. He argued that the institute did not have credibility as a "national conservation education center," pointed out that it had not made any strides toward the development of a conservation education curriculum, and critiqued its failure to affiliate with major universities in the region. Moreover, because the institute's education program was "largely the product of Dr. Brandwein's fertile mind and abilities," his resignation made it unlikely that the institute could carry on his work. Smarting from the implication that he played second fiddle to Brandwein's guiding genius, Brennan shot back that as important as Brandwein had been, the two men had been a team: the institute was not "dependent on any one person."[24]

Train was not yet done. He informed the board of governors that the Conservation Foundation board had met the day before and had agreed to fund a stand-alone task force to investigate the future of conservation education, which his members believed should be focused "on *environmental education* rather than strictly conservation." With this shift in nomenclature, Train uncoupled his institution from the very heritage that moments before Forest Service Chief Cliff had espoused so passionately. In aligning himself with those who were beginning to call themselves environmentalists, as a counter to what they considered old-school conservationists of the Gifford Pinchot era and stripe, Train seemed to join cause with those in the postwar era focusing not on efficient resource management but on quality of life issues, a process that the historian Samuel Hays described as an "innovation of values." Train effectively spurned Cliff's olive branch.[25]

Confirming the linguistic division that Train had articulated, the chief forester replied in kind: the Pinchot Institute "must be strongly identified with conservation," Cliff declared, and must reject the alternatives that had been mentioned, for they "are rather far removed from its central mission." Although the Forest Service later would incorporate "environment" into the name of the next iteration of the Pinchot Institute, during this highly charged in-house discussion the two leaders were drawing bright lines to clarify what was proving to be a set of irrevocable differences. "Recognizing the Conservation Foundation's clear right to reconsider its role in the

Institute," Cliff asked only that the foundation "reach a decision as soon as possible," a request whose foreordained conclusion Gifford Bryce Pinchot then awkwardly announced: "If I were in the Conservation Foundation, I would not think this place has anything particular to offer."[26]

The Forest Service and the Conservation Foundation would paper over their differences in a June 1968 make-nice press release indicating that the two entities were parting company. After "some very serious discussions this year about the Foundation's relationship to the Pinchot Institute for Conservation Studies," the foundation and federal agency "mutually concluded that the cooperative agreement, under which the Pinchot Institute has been operated, should be terminated June 30, 1968." The foundation promised to continue working with the institute "as part of its continuing concern with conservation and environmental education" and to "maintain a close cooperative relationship with the Forest Service on matters of mutual interest." The Forest Service would take on full responsibility for the Pinchot Institute "as a point of national interest, as a center for advanced studies and conferences in conservation education, and a leadership center for a model outdoor classrooms program." But as this letter's signatories, Cliff and Train could only muster the thinnest of praise for the results of their joint venture: "Worthwhile things have been accomplished during the last four years."[27]

With that, Ruth Pinchot finally received the definitive answer to her provocative query in the spring of 1964: "Whatever happened to the Pinchot Institute for Conservation Studies?" It had withered.

Its memory then faded away. Russell Train, for instance, did not mention his role in the Pinchot Institute's final years in his memoir or speak of it in an oral history interview that the Forest History Society conducted with him in 1999. The trail was so cold by the early twenty-first century that his biographer, E. Brooks Flippen, did not pick it up. Educational specialists at the Forest Service have also forgotten the institute's pioneering role in galvanizing agency-wide interest in conservation education. Evidence for this internal amnesia surfaced in a 1999 agency report of the history of its conservation educational initiatives. It contains not a single mention of the Pinchot Institute's creation or its educative mission. Instead, its authors backdate the Forest Service's interest in conservation education only to 1968, when Chief Edward F. Cliff formally established the Forest Service environmental education program. "He was convinced that wise use of the Nation's natural resources would only be possible if the public was educated about these resources," and so he established an "environmental education training team to teach others both the principles of environmental education and how to teach them to others." It was as if the Grey Towers experiment had never occurred, as if the Pinchot Institute had never existed.[28]

CHAPTER EIGHT

Branching Out

All the pleasure of sharing your yard with wildlife builds slowly toward understanding not only wildlife, but man—how we, too, live and fare in relationship to the earth that supplies us with cover, food, water, and living space.

—Jack Ward Thomas and Richard M. DeGraaf,
"Raccoons on the Roof"

A funny thing happened when Jack Ward Thomas and Ronald A. Dixon decided to eat lunch in a graveyard. The two scientists were working in the U.S. Forest Service's research unit at the University of Massachusetts in Amherst and, in hopes of stretching their legs one noon hour, they strolled over to West Cemetery in the town's center: "As country boys, it was not unusual for us to try to escape our stuffy office . . . in search of a more refreshing site to eat." As genial as their environs and meal may have proved, their curiosity got the better of them. They could not help but notice that they were not alone in the bucolic setting: "secretaries, bank tellers, and sales clerks could be seen . . . quietly conversing." Kids on their way to and from school used its winding paths to shorten their route, and "tourists regularly visited the gravesite of Emily Dickinson." Busy too was a surprising variety of wildlife: birds winged overhead, perched on branches and telephone wires, and sang from clumps of bushes. Slipping in and out of dense brush was a number of rodents, while ducks and other aquatic animals laid claim to nearby water sources. There was a vibrant hum of life in this solemn space dedicated to the dearly departed.[1]

Wondering just how widespread this situation might be, and whether it might bear implications for the northeastern corridor of the United States that was desperate for open space, the pair of Forest Service researchers launched a full-fledged scientific study of the kinds of recreational uses and wildlife habitat existing in graveyards located in the Greater Boston area. Over the summer of 1972, they spent two hundred hours interviewing graveyard managers, pouring over land-use maps, and traversing upward of two hundred miles along a transect line. Out of their efforts came a wealth of data. Cemeteries in the metro region accounted for roughly 35 percent of the available open space, making these acres "an extremely valuable resource." Their value became even more clear as the two men recorded the number of bird species (95) and tabulated their reproductive behaviors ("1,195 nests of 34 different bird species," among them starling, robins, and blue jays; ring-necked pheasants and yellow-shafted flickers). They sighted plenty of mammals, including raccoons, skunks, squirrels, and foxes; amphibians, reptiles, fish, and insects were omnipresent. The variety and density of their presence was startling: "Imagine standing in a 3-acre cemetery in south-central Boston, surrounded by 5- to 12-story buildings, and seeing a sparrow hawk deliver a meal of fresh-killed rodent to its young; or walking through a 15-acre necropolis in uptown Brookline and flushing a ring-necked pheasant and her four-week-old brood."[2]

Excited by the capacity of these birds to build a life in what this pair of wildlife biologists called "inadvertent habitat pockets," they wondered whether a "child from the inner city" might feel similarly excited about stumbling upon such "outdoor adventures." First, though, they had to determine how humans made use of these landscapes. That they witnessed more than 1,400 people visit tombs and markers during the survey period was not surprising; it was a bit more unexpected that hundreds walked, jogged, and bicycled through these greenswards. Others fished, bird-watched, flew model planes, trapped chipmunks, dealt and consumed drugs, and played baseball; in one instance, Thomas and Dixon watched a Peeping Tom peep. The intensity and variety of uses, when combined with the rampant vandalism they witnessed, indicated that these sacred terrains, for better and worse, were integrated into the daily life of the surrounding communities. They would be increasingly so in the future, the scientists predicted, for "their green vegetation and towering trees amid urban cacophony and concrete will prove irresistible to people seeking a brief respite."[3]

Yet managing these landscapes for their multiple uses without violating the cultural norms about their significance, which have been "steeped in centuries of tradition and bounded in law," would prove difficult given projected population growth within the already dense Boston-to-Washington megalopolis. For Thomas and Dixon, it was essential to plan for these building pressures: "The space crunch is at hand and time is short."[4]

As innovative as their foray into cemetery ecology was, and as indebted as their framing arguments were to the concerns that a legion of contemporary social critics—among them Samuel H. Ordway Jr. and Fairfield Osborn who helped launch the Pinchot Institute—had advanced about the impact runaway urbanization was having on people and the places they inhabited, what is particularly striking about Thomas and Dixon's research is that they conducted it at all. What were these wildlife specialists, whose employer managed vast swaths of forest and grassland, doing looking at such an odd niche? Why were these Forest Service researchers exploring the interaction between humans, avian nesting behavior, and local burial grounds, while their agency peers, driven by a series of endangered species initiatives, were focusing on recovering bald eagle populations, stabilizing salmon runs, or protecting large mammals? Why urban not rural; East not West? What, in short, were Thomas and Dixon doing in all those cemeteries?

The answer lies partly in the collapse of the collaborative relationship between the Conservation Foundation and the Forest Service in 1967, the federal agency's subsequent reevaluation of where the Pinchot Institute best fit within the organization's bureaucracy, its absorption of the costs associated with managing Grey Towers and its programming, and the reassignment of employees to advance the institute's revised mission. Out of these internal deliberations, which took nearly two years to complete, a new entity emerged known as the Pinchot Institute of Environmental Forestry Research (PIEFR); Thomas and Dixon were among its personnel.

The agency's reconceptualization of the Pinchot Institute was also triggered by the explosive impact that the long, hot summers of the 1960s had on U.S. political culture. As Harlem and Watts, Newark, Philadelphia, Detroit, Memphis, Wilmington, and Chicago erupted in race riots, a fury that was compounded by the back-to-back assassinations in the spring of 1968 of Reverend Dr. Martin Luther King Jr. and Senator Robert Kennedy, and as the furor escalated in response to the country's continuing involvement in the Vietnam War, the executive branch and Congress turned their attention to questions of racial inequity, increasing unemployment, and imploding cities. Civil rights legislation, bills that created the Job Corps and the Youth Conservation Corps, and others devoted to enhancing water quality, decreasing air pollution, and reconstructing streets, parks, and other infrastructure were aimed at resolving some of the key pressures facing urban Americans. In every sense, cities were hot.

Even the Forest Service was drawn into this political ferment. In the late 1960s the vast majority of the agency's personnel and budget were centered on the lands it managed well west of the Appalachian Mountains, and even its eastern researchers and staff were "oriented to the needs of people who intensively managed rural woodlands." That being so, it still could not ignore the signals emanating from the White House or the cues that congressional budgetary allocations were sending. Every session of Congress that debated jobs bills meant that the relevant committees would ask the Forest Service to supply lists of projected projects that might employ Job Corps and Youth Conservation Corps personnel. In testimony before the Senate Subcommittee on Employment and Manpower in June 1960, for instance, Agriculture Secretary Orville Freeman laid out the Forest Service's estimates for $1 billion worth of work in reforestation, recreation development, wildlife habitat improvements, forest fire protections, and road and trail construction that urban youth might be trained to perform.[5]

The agency was just as quick to commit itself to researching and analyzing problems specific to the urban environment, a terrain with which it had decidedly little experience. Yet to secure that knowledge required a shift in priorities and an expansion of responsibilities. One who worked to bring about these necessary transformations was Warren T. Doolittle, the assistant director of the Northeastern Forest Experiment Station, whose offices were located outside Philadelphia and thus situated within the nation's most populous region. It was from this position and place that he pushed the Forest Service to take seriously the challenges that came from urban growth and development. As he put it in a paper delivered at the 1968 annual meeting of the Society of American Foresters—held in the City of Brotherly Love twelve months after it had blown up in an ugly race riot—foresters must realize that the Eastern Seaboard is "becoming more and more vulnerable to pollution and to destruction of natural beauty."[6]

To mitigate these damages would require the agency, in partnership with state and local governments, universities and nonprofit organizations, to develop a new research agenda focused on urban forestry. That was easier said than done, Doolittle conceded. Few agency scientists—few scientists generally—knew much about how urban ecosystems functioned. To discuss whether trees and other vegetation might contribute to healthier cityscapes required some basic information, such as how plants grew under the "exceedingly adverse conditions of soil, water, and air" prevalent in these heavily populated and industrialized areas. If some species were better suited than others to offer sound abatement and light attenuation—how so? If trees can clean the air, cleanse water sources, and cool heated streetscapes, what were their varied capacities to do so? Unknown too was how to protect these woodlands—whether intact forests, local parks, or individual trees—from "insects, diseases, and pollutants."[7]

In response to these challenges, Doolittle announced that the Forest Service, which was "in the thick of the battle against further environmental degradation," was reorganizing some of its program agendas. Its State and Private Forestry branch would be the conduit for matching federal funds to nurture state agencies' efforts in urban reforestation, and its Research branch would use federal dollars to leverage additional contributions from universities and other funding sources to expand the collective understanding of how cities and natural systems interacted. The Northeastern Forest Experiment Station was going to take the lead on this initiative, and its catalyzing agent would be PIEFR.[8]

PIEFR would not be officially established for another eighteen months. In the interim the Forest Service hosted a February 1970 conference at Grey Towers to bring together interested agency personnel, along with representatives of ten northeastern universities with forestry or natural resource expertise, to determine if there was sufficient interest for creating a collaborative relationship between the Forest Service and the academy. There was, and a working group began to draft a consortial agreement. Even before the Grey Towers meeting, the Forest Service had sought budgetary support for this idea from Congress. A key player in these negotiations was Rep. Silvio Conte (R-MA), an avid outdoorsman and conservationist. It helped that Silvio sat on the powerful House Appropriations Committee: over the years he had funneled millions of dollars to the National Institutes of Health for public health research and to the Fish and Wildlife Service to protect Atlantic Salmon riparian habitat (one outcome of which is the Silvio Conte National Fish and Wildlife Refuge on his beloved Connecticut River). Another of the recipients of federal largesse was the five-college consortium located in his western Massachusetts district, which counted among its members the University of Massachusetts. Its forestry faculty, one of whom had attended the Grey Towers gathering, proposed to Conte that he locate funding to support the Forest Service's newfound urban environmental research program. The news that the agency was targeting the Northeast moved the Massachusetts congressman into action. Because of his committee assignment, he had more than enough clout to secure within months a $325,000 "add on" to the Forest Service budget. Half of this new money earmark was designated for the creation of an urban forestry unit at UMass, and the rest would be divided among the consortium as research grants. Conte's legislative maneuver led the Forest Service to transfer Jack Ward Thomas from its research office in Morgantown, West Virginia, to run the new shop and to bring in Richard Dixon and other natural and social scientists to bolster its staff and develop its agenda. Indirectly, Conte made it possible that one day a couple of hungry country boys would head out for lunch in an Amherst graveyard.[9]

Although Thomas and Dixon's resulting research was serendipitous, that was not the case of most of the work conducted under PIEFR's aegis. Its consortial charter argued, for instance, that because the environmental issues facing the Northeast were so complicated, the development of a "cohesive, coordinated research effort is too large for a single university or agency," and studying "local complexity" was "poorly suited to a national institute." Its signatories had concluded that the "best attack" was via "a concerted effort by a regional association of agencies and institutions capable of conducting significant investigations on a coordinated basis." Together, the Forest Service and its partners—which included the universities of New Hampshire, Massachusetts, and Connecticut; Yale, Princeton, Cornell, and Rutgers; Penn State and Syracuse—would "develop the knowledge and technology needed to solve problems of policy formulation, regional planning, and land management related to environmental forestry in and around eastern population centers." Theirs was to be a big canvas.[10]

Without painting themselves too far into a corner, the partners also agreed to prioritize subjects most in need of concerted attention. Securing data necessary to improve environmental planning and decision making topped the list, followed by identifying and increasing the amenities that forests provided metropolitan areas, such as the enhancement of watershed protection and water quality, "wildlife habitat for spectator enjoyment," and the "social wellbeing of urban people through recreation and aesthetics in a forest environment." Sensitive management of the urban environment would also require studies that aided the protection of "high-value forest vegetation from destructive actions of man and other agents," one of the most pressing of which was mitigating the bulldozed clearing of forest cover to construct interstate and intra-urban highway systems. Central to the consortium's work, then, was the advancement of "urban man's understanding of his interrelationships with, and determining his needs for, urban forest environments."[11]

That initial outline for the institute's work intensified with the subsequent development of nine working groups whose steering committees developed research agendas to cover a range of subjects, from urban amenities, tree genetics, air quality, and social and behavioral issues to water quality, wildlife, insects and diseases, land planning and management, and soils. This work required a methodical approach and a motivating rhetoric that came bundled together in a multiauthored report, *The Pinchot Institute System for Environmental Forestry Studies.* Its first words were galvanic, if overwrought: "Like the Four Horsemen of the Apocalypse—harbingers of war, pestilence, famine, and death—four great threats today confront the ecology of the densely populated Megalopolis of the Northeast: water pollution, air pollution, soil erosion, and the destruction of flora-fauna relationships." Its authors challenged environmental scientists to acknowledge that they have been "playing in a kind of three-dimensional chess game" but that they had been concentrating their "efforts on only one dimension of the playing surface." To meet the "real needs of decision makers" would require "recognizing the interconnectedness of the problems, the ramifications of the solutions, and the need to prescribe comprehensive research solutions." Figuring out the "interrelationships within the system" was no small task and came with this important caveat: "This is a first-generation system," and, "[in] much the same way that research on transistors and microminiaturization was required to advance computer technology to second- and third-generation systems," the Pinchot Institute's approach ultimately would have to be tweaked and refined. This would allow it to avoid the usual trap in which public agencies, academic scientists, and think-tank researchers sit within their separate silos narrowly studying only "the bits and pieces of the complex influences, interactions, and contributions of forest resources to the human environment" and fail to develop a coordinated and comprehensive approach to stave off the perceived apocalypse. Systems analysis demanded systematic thinking.[12]

These grand claims notwithstanding, even this initial cohort of scientists had to become comfortable with the new location of their work, their job titles, and the professional nomenclature that in time they would claim as their own. "We're not originally from this part of the country," admitted forest economist Brian Payne, "and until recently we didn't call ourselves environmental researchers." No one did, a fact that Warren Doolittle later recalled was part of the paradigm shift taking place within the Forest Service and the larger culture: "We used the term 'environmental' [forestry] for a number of reasons, partially I think to get into urban forestry and also to use it in a management context. This is when the term environmental was just entering our language and so we wanted to sort of 'catch up' and integrate new thinking into our forest research. . . . The language was difficult at this time and we had to take risks—this was my job."[13]

Just as uncomfortable was the Amherst unit's official mouthful of a title: Improving the Human Environment with Forest Vegetation in the Urban-Forest Interface of the Northeast. Whatever that meant, its personnel was going to have to get used to working across disciplinary boundaries: "The people who put it together felt, quite correctly, that the problems of environmental forestry are too big for any one man or for any one profession to solve alone." This orientation led Forest Service leadership, particularly Warren Doolittle, who in late 1970 was promoted to director of the Northeastern Forest Experiment Station, to staff the UMass office with a wildlife biologist, forest economist, landscape architect, sociologist, and plant geneticist. There were other intellectual boundaries these researchers would need to negotiate, for "conflicts exist among those who manage trees and forest land." Since the late nineteenth century, arborists and foresters had tussled over their professional turf. Local, state, and federal agencies had wrangled over funding, expertise, and priority. Getting beyond these enduring tensions, PIEFR scientists argued, might be as easy as admitting that "the field is large enough and the problems critical enough that all of us, working together, will be none to many." If that approach did not do the trick, a bit of humility might: "As you can see, we haven't done much," Payne and Thomas confessed at a gathering of Massachusetts arborists; "we need your help; there is no doubt about that. And we hope we can be of some help to you."[14]

Over the next fifteen years, the aid that the Pinchot Institute offered was considerable. The official history of the Northeastern Forest Experiment Station encapsulates its achievements in shorthand form: "more than $2.5 million were granted for research studies, problem analyses, symposia, and technology transfer. The Consortium sponsored 11 major symposia and generated more than 300 scientific papers, reports, articles, and presentations, many of which were landmark accomplishments in urban forestry." This enumerated claim is true enough, but it does not begin to delineate the range of work initiated, the context in which it was developed, the legacy it left, or the critiques it generated. Its research—theoretical and applied—was not only important but also instructive of the strategies that scholars, public officials, and activists were seeking to employ and some of the dilemmas they hoped to overcome.[15]

Consider water. Although its flow long had been a focus of Forest Service research, most especially as it related to the protection and management of high-country watersheds, the agency had paid no attention to the problems of urban water pollution, bacterial or industrial. Understandably these metropolitan concerns had been the purview of local and state governments, and these political entities since the Progressive Era had been enacting increasingly rigorous ordinances and legislation to ensure greater public health. Their ability to underwrite such work had declined sharply during the Great Depression, and with the advent of the New Deal, the federal government took up some of the funding slack. Through the Public Works Administration and the Works Progress Administration nearly $500 million had been spent on water and sewer projects. This aggressive posture continued through the 1950s and 1960s: Congress and the Eisenhower, Kennedy, and Johnson administrations pursued a number of legislative fixes and funding mechanisms to increase the provision of clean drinking water, the restoration of polluted streams and rivers, and the treatment of wastewater. As water pollution bills worked their way through the legislative process, and new bureaus such as the Federal Water Pollution Control Administration were created within the executive branch, Washington was signaling that its engagement on these issues was intensifying and that its reliance solely on local governance was on the wane.[16]

The Pinchot Institute's commitment to the funding of water pollution research was a reflection of the federal government's new posture, though its cooperative approach to this endeavor—the distribution of grants to academic institutions, the collaboration between agency scientists and those in other public entities and private institutions, and the commitment to rapidly disseminating their findings—was an important counter to the emerging federal consensus that top-down, mandatory compliance standards would clean up the nation's polluted waters.

Underscoring this point is the 1972 conference that the institute sponsored, "Recycled Treated Wastewater and Sludge through Forest and Cropland." Not exactly a barn-burner of an issue, and certainly not one the Forest Service had ever before supported or studied, the conference ranged across such topics as the biological and chemical qualities of treated effluent and sludge, how soils functioned as filters and how they react to wastewater, and how different ecosystems react to waste disposal. The fact that these subjects were so far afield from the Forest Service's research agenda and historic mission is evidence of its willingness to stretch.[17]

They are a reflection as well of the agency's willingness adopt a mien different from the one too frequently on display during the 1970s controversy over clear-cutting in the national forests. In that volatile context the Forest Service acted on the assumption that its expertise should decide policy, a stiff-arming of the public that damaged its once-vaunted reputation. "I was stunned at how stubbornly the Forest Service pursued its course of action," recalled Jack Ward Thomas, who had been immersed in the battle over timber harvests in the Monongahela National Forest in West Virginia: "We'd made up our minds that even-aged management was the way to practice forestry in those particular circumstances. By God, that was our story, and we were sticking to it. We stuck to it even when it was increasingly clear we were headed for deep, deep trouble. Our response was not to educate, and it wasn't to bring people along. Our response was just to say look, we're the guys that know best and that's what we're going to do. I think that was a huge turning point for the Forest Service."[18]

Other transitional moments, however small, came via the Pinchot Consortium's organizing of the Metropolitan Tree Improvement Alliance (METRIA). Pulling together academics, Forest Service personnel, and arborists and landscapers, the new group emerged after two years of discussions that the Pinchot Consortium and its partners facilitated and funded. As with the wastewater project, METRIA's ambition was to "integrate several disciplines that previously had been scattered and not readily accessible," and likewise to bring together people across the Northeast who had "related responsibilities and interests, but whose paths seldom had met." Each of these projects, then, operated on the assumption that the federal agency and its staff did not know what they needed to know. As the organizers of the wastewater conference acknowledged, the agency's "best bet was to bring together a big group of practitioners and listen to what they had to say."

Learning to listen: in important ways, the Pinchot Consortium was opening up what to the agency's growing chorus of critics otherwise seemed a closed-off Forest Service.[19]

Knocking another hole in its insularity was the Pinchot Consortium–sponsored "symposium-fair" framed around the interactions between children, nature, and the urban environment. Its organizers tested the agency's comfort zone during the five-day event held in May 1975 in Washington, D.C. "Early in the planning process, the program committee agreed not to hold a conference that was within the bailiwick of any single discipline. We were frankly exploring an area of interest," they later wrote, "one that we deemed important, yet one without sideboards established by the conventional wisdom of an established profession." Because they sited it at George Washington University, so as to maximize attendance from public officials, and stacked its participants with a number of celebrity academics, including geographer Yi-Fu Tuan, anthropologist Margaret Mead, ecologist Paul Shepard, and psychologists Rachel and Stephen Kaplan, they drew more than five hundred attendees from nine countries to the proceedings. There they debated the nature of nature, dissected environmental education inside and outside the classroom, and critiqued K through 12 environmental studies. Literary scholars, open space advocates, and urban foresters were part of the intellectual jumble: "We wanted to hear from people doing research as well as from people doing things. We wanted to learn of the dreams of designers. And above all, we wanted interested people—adults and children—to meet together in a pleasant environment, to exchange ideas, share accomplishments, and ask questions."[20]

Such as the challenge that Professor Tuan posed early on about how children absorb the world around them. They are not little adults: "Experience depends on sensory equipment. A child is finely equipped: his senses are sharp, undulled by age. But the ability to make use of his senses is limited," lacking the critical element of time in place and the memories of that an enduring presence. "Remembrance is an important component of appreciation," and building that insight into how humans navigate the urban landscape and guiding, through formal and informal ways, America's youth to a deeper understanding of the terrain they inhabit became key themes of the conference. Determining how children make use of recreational space, how they read their way into a literary nature, discovering what appears to be their innate and learned responses to woods, meadows, and backyards, and the role that television might play in shaping their perceptions of the great outdoors—collectively the presenters were mapping the necessary points of contact that could make a metropolis habitable. "Cities are as livable as the people in them are sane and mature," asserted Pitzer College's Paul Shepard, but the "journey into ecological maturity does not require continuous immersion in a garden." Admitting that it is "a general and probably valid intuition that the destruction of the natural world somehow impairs our humanity," he suspected that "the amount of nature necessary may be surprisingly modest if we can recover the sense of timing and purpose in which it makes us human."[21]

None of these discussions sound like usual federal forestry fare, which was the consortium's point in assembling the conference. It gave a gentle shake to the Forest Service's less-than-consistent appreciation for the need for conservation education, whether rural or urban, for youths or adults. The agency remained enamored of its iconic mascots: Smokey the Bear (1945), the poster-child of the agency's educational outreach about wildfire preventative measures; Lassie—Forest Ranger (CBS: 1964–1971), the television hit that brought weekly lessons in conservation to millions of viewers; and Woodsy Owl (Give a Hoot! Don't Pollute!), which was launched in time for the first Earth Day (1970). These public activities were not met with robust programming from within the agency. In 1961, for instance, it hired the educator Matthew Brennan to join the two person staff of the relatively new conservation education position in the Office of Information. Two years later, with the creation of the Pinchot Institute for Conservation Studies, the agency shifted its educational program, and Brennan, to Grey Towers. Taken together, these were mere gestures toward environmental education than a well-funded set of initiatives. Their uncertain future was illustrated in 1970 when Grey Towers and its anemic environmental programming were shifted out of the Office of Information and to the Northeastern Forest Experiment Station, privileging research not education.

The Forest Service was not alone in its inconsistent embrace of conservation education. The Bureau of Outdoor Recreation, established in 1962 and housed in the Department of the Interior, was just as doubtful as to the value of its assigned task. The new bureau, President Kennedy had declared, would serve "as a focal point within the Federal Government for the many activities related to outdoor recreation," coordinating and consulting with executive branch departments and agencies to implement "Federal outdoor recreation policies." Among these policies was education, which its guiding legislation—the Outdoor Recreation Act of 1963—identified as collaborating with a range of institutions "to assist in establishing education programs and activities and to encourage public use and benefits from outdoor education." This language might sound proactive, but the office interpreted this sentence as signifying that the bureau "should not proffer its cooperation unsolicited but should wait to be asked to help by a college or a university." So it waited, and when asked for aid in curricular innovation and the funding to support it, "the Bureau made cautious attempts to answer the first question and responded with a polite 'No' to the second." Its demurral became prescriptive: when the bureau hired Samuel T. Dana, the former dean of the University of Michigan's School of Natural Resources and a member of the original Pinchot Institute board, to consult on its educational role, he concluded that it did not have one. The movement to promote conservation education through the federal government appeared rudderless.[22]

Hoping to provide some of that much-needed direction had been one of the desired outcomes of the 1975 conference on children, nature, and the urban environment. George Moeller noted that "we can no longer expect children to understand their natural world intuitively, without assistance." As a result he and others urged school districts, individual citizens, nonprofit organizations and private foundations, as well as state boards of education and the federal government, to promote a more thorough and engaged teaching of the natural systems in which humans operated: "Efforts of government agencies, private enterprise, and conservation groups cannot succeed in achieving and maintaining a wholesome environment without the firm support and understanding of the citizenry."[23]

Backyards were no less crucial components of the wholesome landscape the Pinchot Consortium would discover, and they were no less meaningful as a rushing river, deep woods, or a tall-grass prairie. A single telephone call to the Forest Service's UMass research unit made this case convincingly. On the line was George H. Harrison, the managing editor of *National Wildlife*, the glossy publication of the National Wildlife Federation. The federation had emerged out of the 1936 North American Wildlife Conference, which staunch conservationist J. Ding Darling, then head of the Biological Survey (and forerunner of the U.S. Fish and Wildlife Service), had convinced President Franklin Roosevelt to organize. Its purpose was to unite the many disparate conservation groups across the country and to give them, and the wildlife they cherished, a stronger voice in the political arena. Although its efforts ever since have focused largely on protecting rural landscapes, watersheds, marshes, and charismatic species, in the late 1960s it began to recognize that urban ecosystems ought to be studied and urban voters cultivated, just as the Forest Service had. "You are interested in ecology and wildlife and a clean environment," *National Wildlife* editor Joe Strohm wrote his readers, "but you and I must recognize that it's harder and harder to sell these ideas to people brought up on concrete." As part of the federation's effort to reach out to this new constituency, Harrison telephoned Jack Ward Thomas.[24]

He knew about Thomas's recently completed dissertation on urban songbird habitats, was aware that the Pinchot Consortium's focus in Amherst was on promoting wildlife habitat within the built landscape, and shared with these scientists the recognition that the "very same basic principles wildlife managers had been using for decades—providing food, water, cover and places to raise young—worked beautifully on a smaller scale in backyards." Harrison proposed that Thomas and his colleagues pull together their findings for his popular magazine's large audience. There was a catch: they had less than a week to produce the article, already slated for publication in the April-May 1973 issue.[25]

Well aware of the unparalleled opportunity Harrison's offer presented to transfer their ideas from the field into daily life, Thomas worked with Robert Bush, the unit's landscape architect who handled the artwork, and Richard DeGraaf, another wildlife biologist, to work out "the technical wildlife and plant materials." Three days later, draft in hand, the trio drove to the Hartford, Connecticut, airport where they met the magazine's layout team. Huddling together in the small private plane parked on the tarmac, they edited the manuscript, tinkered with its design elements, and agreed on its format. "Invite Wildlife to Your Backyard" was the result. Its straightforward prose, clean graphics, and simple message about the capacity of every suburbanite to turn monoculture crabgrass into a "refuge-in-miniature" struck a chord with the public. Many of the contemporary crises, such as those about smog-choked cities and toxic waterways, seemed beyond the capacity of any one person to resolve. But that was not true in a suburban yard or urban window box, where, with some tools, muscle power, and a bit of cash, homeowners and renters could "rub shoulders with nature." Indeed, their individual restoration projects could have a collective impact: the final segment of the article announced a new program "to establish a nation-wide network of mini-refuges in the backyards of Federation members." If adopted wholesale, "vast amounts of land in residential neighborhoods could be turned into a tremendous asset for wildlife—and people."[26]

The concept took off. Within months, an "avalanche of mail" surged into the federation's Washington, D.C., offices. Thousands of people applied for the backyard wildlife program, and government agencies, corporations, local conservation groups, schools, churches, and other civic institutions requested as estimated 250 million reprints. This response fed on itself as the National Wildlife Federation sent press releases to local media every time one of its member's backyards qualified for the program. "We seem to have created a new kind of celebrity," chortled James D. Davis, director of Creative Services at the Wildlife Federation. Senior citizens and children and neighborhoods were being cast in the spotlight "for simply doing their thing as good conservationists. In this cultural era of the 'non-hero,' we don't think this is a bad thing."[27]

The Pinchot Institute had done well too by the federal money channeled through it to bring together scientists, educators, and citizens for a range of public events, research initiatives, and academic conferences. Because these dollars leveraged additional cost-sharing funding from universities, and state and local agencies, the bang for these bucks was increased. In providing "seed money for projects that, once moving, showed sufficient promise for other sources to accept responsibility," the consortium could take considerable pride in its actions, for the result was "a rippling effect whereby small but strategic federal investments totaling almost three million dollars contributed to research beneficial to millions of people living in the urban and suburban areas of the northeast."[28]

Those benefits might have even been more widespread had the consortium scientists inside and outside the Forest Service been more willing to publish in nonacademic journals. The Amherst research unit, which produced the *National Wildlife* backyards article and a series of related pieces designed to reach a lay audience, proved the exception. More often, pathbreaking work in acid rain research, tree-growth dynamics, or air and water quality appeared only in professional venues; admittedly, this is largely the result of the need for pre- and post-tenure faculty to meet the publishing requirements of their home discipline. Still, the original consortial agreement called for dissemination of such information beyond the tight confines of the academy, an ambitious goal rarely met.[29]

Had there been more outreach there might have been more consistent financial support: building a larger constituency for its work might have insulated the consortium for the budgetary retrenchment that occurred during the Carter and Reagan presidencies. These chief executives may not have agreed on much politically, but they concurred that federal spending needed to be reined in. As they shrank expenditures across the 1970s and 1980s, so did the amount available for the Pinchot Consortium to distribute. After reaching a peak of $271,847 in budget year 1975—the year before Carter's election—it dispersed $250,553 in 1980; in 1984, its last year of operations, it released a mere $118,540. Money, and its lack, brought a halt to this innovative experiment in cooperative science making.[30]

As striking was the consortium's insistence that its funding underwrite interdisciplinary research driven by a commitment to making the northeastern urban corridor—the megalopolis, stretching from Boston to Washington, D.C.—more habitable, clean, and safe for rich and poor alike. Over its fifteen-year run, the Pinchot Institute boosted the status of urban forestry as a tool for ameliorating some of the most visible environmental ills that afflicted America's cities, and in the process the institute transformed the lives and careers of those who worked in the field and lab to resolve them. It provided an entrée into the profession for a slew of graduate students at the participating universities. Altered too were the career trajectories of Forest Service personnel, among then Richard DeGraaf and Jack Ward Thomas. They had worked hard to convey how their scientific findings redounded to the public good, an outreach that continued to shape their writing long after they had left Amherst. In time, DeGraaf would rise to become the chief research wildlife biologist for the agency's Northeastern Forest Experiment Station. The Forest Service moved Thomas west, transferring him to its research unit in La Grande, Oregon, which he managed for the next twenty years. One of his key assignments in the early 1990s was to be the head of an interagency team assessing the impact that logging had on Spotted Owl habitat in the Pacific Northwest. That work brought him to the attention of President Bill Clinton, who tapped Thomas to become the thirteenth chief of the Forest Service.[31]

In quieter, more indirect ways the Pinchot Consortium helped ordinary citizens come to terms with the world they inhabited. Moved to act by the step-by-step guide to a healthier backyard ecosystem that *National Wildlife* had published, Carole Mebus and her family converted their Easton, Pennsylvania property into a refuge. "The natural world can be more calming than a tranquilizer and at the same time more invigorating than a stimulant," she later wrote to the magazine, and "whatever the problems of the world, Watergate or the energy crisis, they pale in importance when I hear the call of the wood thrush as it drifts from the woods far away. The wrens dart in and out of their nesting box to the accompaniment of hungry voices, and I feel the world will continue for another year." Out of the spontaneous, productive cooperation between the National Wildlife Federation and the Forest Service had come what Emily Dickinson, Amherst's poet laureate, had defined as "the thing with feathers"—hope.[32]

Grey Towers shortly after construction, 1886. Grey Towers NHS Collection

Forester Gifford Pinchot (*center*, white shirt) on timber-marking operation in Yellowstone Forest Reserve, early 1900s. Grey Towers NHS Collection

Yale Forest School, class of 1906, at Grey Towers camp. Grey Towers NHS
Collection

(Opposite, top) Gifford Pinchot, Mary Eno Pinchot, and James Pinchot, Grey
Towers 1905. Grey Towers NHS Collection

(Opposite, bottom, left) Amos Pinchot. Grey Towers NHS Collection

(Opposite, bottom, right) Ruth Pickering Pinchot. Grey Towers NHS Collection

Gifford Pinchot and Theodore Roosevelt, Inland Waterways Commission,
1907. Grey Towers NHS Collection

Corneila Bryce Pinchot at suffragette parade in New York, 1917. Grey Towers NHS Collection

(Below) Gifford Pinchot's bedroom desk, Grey Towers. Grey Towers NHS Collection

Gifford Pinchot campaigning for governor of Pennsylvania, 1930s. Grey Towers NHS Collection

First Lady Cornelia Pinchot with Sweatshop Strikers in Pennsylvania, 1930s. Grey Towers NHS Collection

Gifford Bryce Pinchot and Gifford Pinchot doing what they loved best, fishing. Grey Towers NHS Collection

(*Right to left*) Governor William Scranton, Samuel Ordway, President Kennedy, Ed Cliff, Orville Freeman. U.S. Forest Service, Forest History Society Photo Collection, Durham, NC, USA

President John F. Kennedy (*left*) and Gifford Bryce Pinchot (*right*) leave Grey Towers after a tour of the mansion that followed the dedication. Pinchot Institute Dedication Album, Forest History Society, Durham, NC, USA

Pinchot Institute sign at Grey Towers, September 1963. U.S. Forest Service, Forest History Society Photo Collection, Durham, NC, USA

Grey Towers, September 1963. U.S. Forest Service, Forest History Society Photo Collection, Durham, NC, USA

The inaugural board of the Pinchot Institute. Seated (*left to right*): Laurance Rocke-feller, Fairfield Osborn, Edward Cliff, Sam Ordway, Samuel Dana, and David Heyman; standing (*left to right*): Paul Brandwein, George Brewer, Joseph Fisher, Wilson Clark, George Selke, DeAlton Partridge, Richard Droege, and Matthew Brennan. The only member of the board missing was Gifford Bryce Pinchot. Grey Towers NHS Collection

(Opposite, top, left) Grey Towers Director John Gray, 1970s. Grey Towers NHS Collection

(Opposite, top, right) Grey Towers Director Ed Vandermillen, 1980s. Grey Towers NHS Collection

(Opposite, bottom) This terraced clear-cut in the Bitterroot National Forest was among those that prompted Gifford Bryce Pinchot to challenge the Forest Service's timber harvests in his father's name. U.S. Forest Service, Forest History Society Photo Collection, Durham, NC, USA

Gifford Bryce Pinchot, late 1970s. Grey Towers NHS Collection

(Below) Gifford Bryce Pinchot and Sally Pinchot aboard *The Loon*, mid-1980s. Grey Towers NHS Collection

Planting an apple tree in memory of Gifford Bryce Pinchot, 1990.
Back row (*left to right*): Ed Brannon, Mariana Kastner (Gifford Bryce Pinchot's
daughter), Peter Pinchot, Gifford Pinchot III (Gifford Bryce Pinchot's sons), Sally
Pinchot (Gifford Bryce Pinchot's wife), Joel Kastner (son of Mariana), Alex Pinchot (son of Gifford III), Jeremy Pinchot (son of Peter), and Max Peterson (retired
chief of Forest Service). Front row (*left to right*): Leila Pinchot (daughter of Peter),
Marianna Pinchot (daughter of Gifford III), and Ariel Pinchot (daughter of Peter)

Director Edgar Brannon and Paul Labounty inspecting slate roof restoration, late 1990s. Grey Towers NHS Collection

Pinchot Institute President Al Sample and Peter Pinchot at the EcoMadera project, June 2007. Pinchot Institute for Conservation

Restoring a forested riparian buffer along the Paulins Kill will capture runoff and cool the stream. Pinchot Institute for Conservation

Landowner Gary Carr discusses forest management activities underwritten by the Common Waters Fund. Pinchot Institute for Conservation

CHAPTER NINE

Turning a White Elephant Gold

It is to be hoped that Grey Towers may have in the future, through instruction and original investigation, . . . a wide and continuing influence in all matters relating to forestry.

—James Pinchot, 1907

You can almost hear the fatigue in his voice. As Gifford Bryce Pinchot read through the 1982 master plan for Grey Towers, an environmental and cultural assessment that the National Environmental Policy Act (1970) required in advance of any sustained alteration to the national landmark, he was unimpressed and saddened. He had seen such documents before, having received any number of Forest Service plans over the past two decades promising the revitalization of his family's former home. Tired of the agency's failure to execute on previous ideas for how to best utilize the Milford manse, he was tired too of his role as curmudgeon in calling out its repeated failings: "I've . . . made some attempts through various chiefs to get some really constructive use of the place," he wrote John L. Gray, who had become director of Grey Towers in 1978, but "I don't feel I was very successful."[1]

Pinchot's weariness with and wariness of the process did not stop him from once more sending a terse letter to Gray. Among other things, he was puzzled why, after seven years of meetings to plan the plan, no one in the agency had approached him about one of its key components—the purchase of adjacent acreage that the family owned so that the Forest Service might establish a visitor center and demonstration forest as part of the site's ramped-up environmental education programming. "I do want to make it clear that the Forest Service doesn't have any commitment from me to transfer more land to it," he wrote. Observing that the property in question was held in trust for his children and his aunt Ruth Pinchot, and that as a result it would take "a rather complicated legal procedure before anything could be done," he was skeptical about pursuing such a path in any event: "before undertaking anything like this, I would want to be assured that such land would be put to more productive and sensitive use than Grey Towers [has been] under Forest Service management."[2]

Pinchot had considerable doubts about the land-management agency's capacity to restore the mansion appropriately, sensitively. This new plan, for example, gave no indication as to what era the restoration would be pegged: When his father was Forest Service chief (1905–1910)? When his mother renovated Grey Towers in the 1920s and 1930s? When the house was transferred to the Forest Service in 1963? Not being explicit about this time frame meant that the planning group had neither understood the basic requirements of historic preservation, and the federal laws that guided any such work, nor had it given serious consideration to the extent, nature, or cost of the rehabbing project whatever the intended historic restoration. Besides, budgetary pressures had derailed every earlier proposal—"as you are perhaps aware, a joint study by the Conservation Foundation and the Forest Service was made some years ago . . . [and] it was the conclusion of this study that so much of the original furnishings and artwork and plantings had either been lost or destroyed, that restoration was virtually impossible"—leaving Pinchot to conclude that the same thing would happen to this one.[3]

Complicating the Forest Service's aspiration was the fact that Grey Towers "was a unique expression of the taste, imagination, and artistic sense of my mother and some famous architects and artists," whose flair would be difficult for anyone to replicate. The agency had shown in the past that it did not understand this challenge: the last historical architect it had hired to draft restorative plans "wasn't interested enough to come here to see pictures of all the rooms in the house as they were when the house was transferred to the Conservation Foundation and the Forest Service."[4]

Even as he apologized to Gray for writing "such a negative letter," Pinchot could not help but reflect on why he had felt compelled to do so: "I have been somewhat shocked by the destruction that took place at Milford under the Forest Service's management." Every bright idea either missed the point of Grey Towers or ignored its potential, an ongoing failure he chalked up to the agency being an agency: "I would like to see the place restored, as I think it was a unique expression of a very imaginative person's taste. Somehow, I don't see this being done through the usual bureaucratic channels, and I think if it is to be done at all, someone with very unusual taste and understanding has to be put in charge and given a free hand." He would not live to see that welcome change; three years after his death in 1987, Edgar Brannon was tapped to be the director of Grey Towers, and he quickly launched a ten-year, $20 million restoration project that when completed in 2000 finally made the national landmark whole.

Until then, however, the once-grand country estate languished. Indeed, when faced with an opportunity in the late 1960s to address the already established pattern of neglect and indecision about the building and related programming, the Forest Service had failed to act. Stung by the collapse of its partnership with the Conservation Foundation and uncertain about what it should do with the mansion, its tiny staff, and the site's mission of promoting conservation education, the agency's leadership made two critical decisions. In 1968 it discontinued conservation programming at Grey Towers, and Matthew J. Brennan, who had been the on-site director of the site, left its employ. At the same time, Grey Towers was shifted out of the Division of Information and Education in the Washington Office and to the Research branch under the new direction of its Northeastern Forest Experiment Station in Broomhall, Pennsylvania. These bureaucratic reshufflings were paired with another significant resolution: the agency determined that Grey Towers no longer needed a resident director and so replaced Brennan with John P. Walsh, who held the title of administrative assistant. The place and its personnel had been demoted.[5]

Demoted and set aside—that's the gist of an internal 1970 memo that discusses the transition envisioned for Grey Towers: "When Research took on Grey Towers, we could see important, intensive, and constructive use of the building in the long run beginning in about ten years." This delay confirmed Gifford Bryce Pinchot's worst-case reading of the diminished status of Grey Towers. Had he known, he hardly would have been pleased that the memo also acknowledged that the Northeastern Forest Experiment Station's leaders had been "negligent in keeping him [Pinchot] informed."[6]

It wrote the site out of its public announcements about the 1970 establishment of the Pinchot Institute for Environmental Forestry too. Forest

Service Chief Edward Cliff's major address that August about the new entity contained but a single sentence about Grey Towers: "opportunities for environmental-forestry demonstrations and education at the Pinchot Estate will be enhanced by the Institute." Somewhat more forthcoming about the significant changes that were afoot, and the resulting downgrading of the national landmark, was Warren Doolittle at the Northeastern Forest Experiment Station. Noting that the "conservation studies" aspect of the Pinchot Institute has been replaced by "environmental forestry," and admitting that "little to no research will be conducted on the Pinchot Estate itself," he assured Pennsylvania foresters, many of whom had taken a keen interest in the original plans for Grey Towers, that "a program of demonstrations, information, and education" will remain in Milford; the Pinchot mansion will continue to function seasonally "as a center of Pinchot history—as it has been for the past seven years."[7]

With that, Grey Towers disappeared from view. Its budget was capped at a little more than $100,000 a year, the preponderance of which was dedicated to staffing and routine maintenance. During the summers, a small cadre of volunteer docents led abbreviated tours of the mansion's interior and grounds; what the eight to ten thousand visitors encountered inside was a quite modest set of displays. Hoping to develop a more substantial experience for them, in March 1972 the Forest Service consulted with the Smithsonian Institution on how better to display the few Gifford Pinchot–era artifacts that remained. In his subsequent report, William Hasse, assistant chief of its design division, cautioned Walsh and Doolittle that any object that remained in rooms lacking air conditioning or humidity control would deteriorate; while "lighting and climate controls seem quite adequate for the cooler darker months," that was not the case in the hotter, brighter, and more humid summers, putting the historic furnishings in jeopardy. Not until the early 1980s was a full-time curatorial staff hired to begin to implement a professionally designed set of exhibits, and it was not until the mid-1990s that the HVAC system was fully upgraded. These delays tell a larger story about the impact of the Forest Service's "deliberate" decision not to enhance the site or bolster its contributions to the agency's land-management mission.[8]

The first sustained effort to clarify Grey Towers' status came in November 1975, when, prompted by an inquiry from local Congressman Joseph McDade, a high-level Forest Service planning group met at Grey Towers to devise a series of alternative futures for the site. These options then were presented to the agency's top leadership—excepting Chief John R. Mc-Guire—at a February 1976 gathering in Milford, and they proved to be the most significant dialogue about the estate and its status within the Forest Service since 1963.

Everything was on the table at that session, including terminating the agency's ownership of Grey Towers. In addition to its disposal, there were more positive choices, ranging from maintaining the status quo, developing a modest increase in programming and related budgetary outlays, defining and funding a more substantial upgrade, or utilizing the site to its maximum potential. The intense give-and-take around the conference table lasted more than an hour, and because this discussion provides a fascinating window into how the Forest Service's upper echelon grappled with its responsibilities for the agency's history, the mansion's legacy, and the present-day dilemmas that Grey Towers illuminated, it seems appropriate to let it unfold largely as the stenographer captured it:[9]

> *Philip Thornton* (Deputy Chief, State and Private Forestry): We have not met our previous commitments to Grey Towers.
>
> *Thomas C. Nelson* (Deputy Chief, National Forest System): Due to the location of most of the Forest Service in the West, I cannot see this as a National Training Center. We do not have enough people in the Northeast to justify a Regional Training Center. We need a cost analysis to compare Grey Towers with other [Forest Service] training centers (housing, per diem, travel, cost of holding meetings).
>
> *Thornton:* There should be a mix of in-service and out-service training, not strictly an FS training center.
>
> *Rexford A. Resler* (Associate Chief): We need to pull in outside groups—AFA [American Forestry Association], SAF [Society of American Foresters], attempt to share costs of operation. We could back out of Grey Towers. Should we toy with this? I do not feel we should unless the funding situation requires it.
>
> [The] FS is in stress in the 1970s—Organic Act, RPA—a State of Challenge. This presents an opportunity to lead off with new initiatives. Enhance our image externally and internally. Use Grey Towers as a center of high-level meetings. We need to pull together other agencies, groups—a melt—knowledge transfer and goodwill. Could give us a shot in the arm; stimulate our own people.
>
> [Grey Towers is] an opportunity we have sloughed off. I'm in a go-for-broke mood. There has been a dearth of leadership, and support

for Grey Towers. A chance to go with a consortium, a chance to go with EE [environmental education] on the East Coast. We should develop and run.

Warren T. Doolittle (Associate Deputy Chief): I have been opposed to [the] FS collecting white elephants. This is different. We would have a little more than "egg on our face" if we backed out.

Nelson: Principle value—historical; research—not a prime location; we tend to be going toward the college campus. YCC [Youth Conservation Corps]—this activity could be held anywhere. Core block—history.

Robert M. Lake (Director of Information): . . . the I&E [Information and Education] people who met here . . . indicated their enthusiasm for an EE program at Grey Towers.

Einar L. Roget (Associate Deputy Chief, Programs and Legislation): Historical values are uppermost. Problems associated with uses that would require traveling here are insurmountable. We should make plans to restore the historical values.

Thornton: We are hanging up on history. Tradition and inspiration values—they are key. Would a motel room in Chicago provide the same setting and mood that we have here at Grey Towers?

Jack Deinema (Deputy Chief of Administration): We should not call it a National Training Center. We should take a cross section of all of our training meetings and hold them here. Far more meetings here, on a deliberate basis.

Thornton: I am not satisfied with our present program. We must do more.

Deinema: Why isn't the operation of Grey Towers financed by overall funding? Why only from P+M [Project and Management]?

Group: Agreed.

Thornton: An expanded program should not be run by research.

Doolittle: Any increase in activity ought to result in administration being turned over to State and Private.

Resler: Not enough support being provided by WO [Washington Office] and the rest of the FS. Need assistance in developing an EE and VIS [visitor information services] program—soon. Before we can develop costs. . . . Emphasis to come from WO.

Roget: Someone should also look at it from a historical aspect; restore [Grey Towers] to original condition.

Nelson: What is EE costing us?

Lake: WO is spending $33,000—$200,000–300,000 being spent Service wide.

Nelson: We must identify EE in budgeting process.

Roget: Whole facility [i.e., Grey Towers] should be a separate budget item.

Lake: An EE training center was proposed here a few years ago, but it was not put in as a line item.

Nelson: Where is FS going in EE? Should HEW head it or FS? Pinchot as Governor of PA should be included in our VIS plan—the whole man.

Deinama: A VIS, EE plan would give us something to work with. General assessment needed to finance this operation. If we had a program to go to [Rep. Joseph] McDade.

Thornton: We are letting the family down. We are letting the FS down, if we go on our present course. What do we tell McDade for FY '77?

Resler: Bob [Lake] I would like you to develop a program. Consortium idea, EE, VIS, Manpower, Hall of Fame, separate or together?

Thornton: A $300,000 package for McDade for FY '77.

Resler: In a month's time we need a general plan.

Thornton: We are not going to get out—we are not going to piddle around. A $300,000 budget minimum. Do it or hang up the jock!

Group: Agreed.

Within a month, Robert Lake had compiled a comprehensive outline, slapped on a grandiose title, the North American Institute for the Future of Forest Environments, and sent it to Associate Chief Resler. The report called for the upgrading of the visitor information services provided at the mansion, "with a strong tie to Pinchot the man, and the conservation ethic and history of Grey Towers." It also boosted Youth Conservation Corps manpower programs that would facilitate maintenance of the house, grounds, and forest and provide labor for Milford and environs, and it promoted an environmental education initiative that would be coupled with an agency-specific training component and conference center. The construction of a Conservation Hall of Fame Trail, "with interpretative signs and benches where the words of great conservationists can be contemplated in a forest environment," rounded out the major offerings.[10]

None of these projects could occur without increased funding from Congress, and in April Resler and Doolittle met with Congressman McDade, a member of the House Appropriations Committee. In McDade the agency had a good friend, much as it had had with Rep. Silvo Conte, another House Appropriations Committee member who had engineered the initial outlays for the creation of the Pinchot Consortium. Like Conte, Rep. McDade took good care of his constituency, directing millions of federal dollars to his district, Pennsylvania's tenth, which he served from 1963 to 1999. An avid supporter of Steamtown National Historic Site in his hometown of Scranton and of nearby Tobyhanna Army Depot, he was a prime force behind the creation of the Delaware Water Gap National Recreation Area (which the National Park Service took over after the government stopped the Tocks Island Dam project). What caught his eye about Grey Towers was that this federal installation was located in economically depressed Pike County. Although McDade had repeatedly pressed the Forest Service to bring him major funding proposals, it had not done so, yet another signal that for the agency the Pinchot Estate was of minor import.[11]

McDade was receptive when Resler and Doolittle brought him the results of the February 1976 conference, with its request for a $500,000 earmark for the 1977 fiscal year. Yet he was not immediately persuaded, believing the agency's plans were too modest. McDade "approved the concepts but asked for more detailed plans and alternatives, including preliminary architectural and engineering plans and cost estimates associated with capital investments to use the site to its full potential." His response set up yet another round of developmental planning and consultations that consumed the next fifteen months. On July 25, 1977, the good news came: Congress had passed a $522,000 appropriation for the operations of the Pinchot Institute for Conservation Studies.[12]

The Forest Service's leadership and Rep. McDade were quite pleased with the results of the preceding thirty months of negotiation. The Pinchot family, however, was skeptical that the proposed institute would ever materialize. After reading through the relevant documents, Gifford Bryce Pinchot and Ruth Pinchot shared their disappointment. "I don't make any more out of this letter than you do," Gifford wrote, and he did not because it sounded exactly "like the thing that they were talking about when they started the Institute [in 1963], but nothing ever seems to get done. I'm sorry not to have more cheerful news, but that seems to be the way things go."[13]

This time, however, the outcome was (largely) different. Grey Towers was pulled out of the Research branch and placed under State and Private Forestry, reporting to the director of its Northeastern Area Office in Upper Darby, Pennsylvania. Its budget increased, and its sourcing was to be made "permanent rather than an add-on." This more substantial and secure outlay allowed the Forest Service to expand staffing. In 1978 John L. Gray, who was serving as interim director of Grey Towers, was appointed as the director, the first to hold that more-elevated status in a decade. Although he had only been in the Forest Service for a year, Gray had had considerable administrative experience; between 1963 and 1977 he was the director of the University of Florida's School of Forestry and is credited with expanding the research focus of the faculty, enhancing undergraduate and graduate curricula, and initiating and supervising the construction of the school's new building. It turns out those details were a significant reason why, on his retirement from academia, the Forest Service was happy to appoint him to the Washington Office as a Research Forester—and why within months Chief John McGuire asked him to serve as interim director of Grey Towers.[14]

Gray's central task was to manage the completion of the site's first full master plan. He also had authority to hire eleven full-time staff, including an assistant director. He was charged with increasing the number of Youth Conservation Corp enrollees on site, and he brought on five interpretative specialists to lead docent tours during the summer to serve an estimated twenty thousand visitors.

No less indicative of the new energy that the McDade appropriation unleashed was the commitment to recovering Grey Towers' architectural history and material culture, as well as the Pinchot family's multigenerational impact on Pennsylvania and the nation. The Forest Service hired outside archivists to sift through Gifford Pinchot's voluminous papers in the Library of Congress to locate documents relevant to these intertwined stories. It consulted with non-agency curators to advise it on how to develop a historically rich set of exhibits at Grey Towers. Identifying just how bad the resources had been managed was only part of the job, one consultant concluded after interviewing unhappy community members and the Pinchot family. His recommendation was blunt: as the agency geared up to preserve the house and its furnishings, it also must mend these frayed social relations.[15]

The first step in that process was the production of an in-depth *Historic Structure Report* of Grey Towers that the Forest Service's chief historian David Clary oversaw. This 1979 document was the first that the agency had ever developed for any of the hundreds of historic sites that existed on the 193 million acres under its stewardship. That lapse is striking. In 1963 Secretary of the Interior Stewart Udall had proclaimed Grey Towers a National Historic Landmark, citing provisions of the National Historic Sites Act of 1935, which declared that "it is a national policy to preserve for public use historic sites, buildings, and objects of national significance." Three years later, with the passage of the National Historic Preservation Act of 1966, more robust regulatory language determined federal agencies' obligations; the Forest Service and peer agencies were required to administer Grey Towers' "historic resources in a spirit of stewardship for the inspiration and benefit of present and future generations." In the words of William J. Murtagh, the first Keeper of the National Register of Historic Places, the new act prioritized the protection of "the cultural ecology of America's built environment," a claim with which the historian David Hamer concurred, arguing that the law marked "the point when the federal government took account of its own responsibilities in the fast unfolding program of historic preservation."[16]

In achieving this lofty goal, the Forest Service lagged behind other federal entities, which concerned chief historian Clary. With the publication and widespread internal distribution of the 1979 Grey Towers historic structures report, he hoped to educate agency leadership and thousands of employees about their neglected duties—to Pinchot's estate and many other sites. The document's real worth, he wrote, is that it illustrates "the kind of investigation, thinking, and decision-making that should be applied to such sites managed by the Forest Service. The instructions this report presents are of the kind urgently needed not only by historic preservation specialists, but by nearly all of our personnel whose activities in one way or another affect these resources." Just as Grey Towers was supposed to embody Gifford Pinchot's lifelong commitment to a dynamic and practical conservation, so this historical analysis of his home would serve a vital heuristic purpose, "as a model for other reports of its kind."[17]

The document offered a well-researched analysis of Gifford Pinchot's long career in public service, recovered the original designs for and construction components of Grey Towers, and enumerated the later renovations that changed its physical form and the family's use of it. In addition to a strikingly candid assessment of the poor state of the existing fabric of Grey Towers and its outbuildings, and the extent of structural restoration required to bring them up to code and back to their historic feel, the report was explicit about the agency's flawed stewardship of the original art collections and furnishings in its care: "Many of the pre-1963 materials are in a serious state of repair caused by improper conservation treatment, visitor abuse, or general neglect. Many of these items require immediate attention if they are to be saved." Reaching out to the Pinchot family could help in their rescue: "In the eleventh hour of the research of this report, it was discovered that Dr. Pinchot retains a remarkable collection of family photographs of Grey Towers, as well as various other materials, from home movies to architectural drawings. Surely the family members are the best source of information on many diverse aspects of the estate's history. They have a genuine interest in fostering Grey Tower's success as a historic site," the report commented, "and a concerted effort should be made to solicit their cooperation." Doing so would also enable the Forest Service to operate at Grey Towers with the same self-proclaimed obligation that defined its management of the national forests and grasslands: to care for the land and serve the people.[18]

But did these necessary reforms amount to what Chief McGuire described as a "new awakening" for Grey Towers? Certainly the landmark had the attention of senior staff during the late 1970s and early 1980s. The local media homed in too on Pike County's so-called Camelot, responding to an uptick in the number of press releases that Gray and his colleagues churned out announcing a range of new environmental education programs and guided tours for a swelling number of visitors. The new script from which the docents worked was not for everyone, and this was by design: "The institute would not be doing its job," Gray told the *Pocono Record,* "unless it's considered a little way out by the conservatives. It should function as Gifford Pinchot would want it to." Also generating press buzz were the Youth Conservation Corps and Young Adult Conservation Corps enrollees who trained and worked at Grey Towers and cleaned up and replanted local parks in Milford and surrounding communities. As for the leadership-training sessions for agency personnel, these sparked a newfound excitement about Grey Towers, not incidentally nurturing service-wide buy-in for the site. The many who over the years attended workshops within the historic facility and hiked its serene, wooded landscape would have happily answered in the negative to the rhetorical question that Philip Thornton had posed to his chief and staff colleagues in 1976: "Would a motel room in Chicago provide the same setting and mood that we have here at Grey Towers?"[19]

Less certain was whether the momentum that came out of that watershed meeting, and that characterized John Gray's four-year tenure at Grey Towers, would be maintained when Gray retired in 1982. The completion of that year's master plan was his swan song, and it contained clues about some of the bureaucratic hindrances that would afflict Grey Tower's operations in the coming years. Most notable was the insistence that all Grey Towers' programming must be integrated with and contribute to the work of the three branches of Forest Service administration: Research, State and Private, and the National Forest System. The rationale for this trifocal lens was bureaucratic: "There have been 4 or 5 plans developed over the years," one contemporary report noted, "none of which were implemented. The probable reason is that none of the former plans reflected a broad program that could assist in accomplishing the three missions of the Forest Service." Because previous iterations did not support the work of one or another of the three divisions, in meetings in which budgets were divvyed up or internal policies were set or hiring levels were established, Grey Towers did not garner a full range of support or attention.[20]

This political analysis of the site's vulnerability also meant that, once corrected, its programmatic energies had to conform to the stated goals of the three administrative divisions. To ensure that its actions matched up with Research, for instance, the corresponding mission of the new Pinchot Institute for Conservation Studies was crafted so that in some unstated way it would serve "as a catalyst to expedite the transfer of . . . technology to the field forester." Its obligation to State and Private Forestry was similarly opaque: Grey Towers "would model new and innovative approaches in the field of urban forestry," though surely that had been and remained the work of the Pinchot Consortium. And despite the presence within the National Forest System administration of a large cadre of policy analysts, the tiny Grey Towers staff would be enlisted to analyze the need for and offer recommendations for "national policy changes and implementation." The implication of these demands on its staff's time was that they might well lead to a narrowing of Grey Towers' work, a limiting of its professional contributions to the agency. Rather than acknowledging Grey Towers as an unusual component in the Forest Service, with a unique opportunity to expand the agency's objectives, those planning its future sought to meld it into the whole, delimiting its sphere of action and constraining its options. In committing to provide services to every element of the agency, Grey Towers would end up doing little to distinguish itself. That was exactly what Gifford Bryce Pinchot had feared would occur if the landmark's future was constructed along the lines of the "usual bureaucratic channels."

Richard McArdle, a former Forest Service chief from 1952 to 1962, voiced similar concerns. He had been involved in the original negotiations with Gifford Bryce Pinchot on the possible transfer of Grey Towers, and during the public-comment period about the 1982 master plan he attended a session devoted to it at the Washington, D.C., headquarters of the American Forestry Association; afterward he wrote a lengthy letter to John Gray about his reactions that reveal how agency leadership calculated which of a variety of choices made the most sense.

About the Grey Towers presentation in the nation's capital, McArdle had nothing but praise to offer: "it is obvious how many of your staff have fallen under the spell of this wonderful old house. I have visited many historic homes across the country. Although each has distinctive qualities, I feel that Grey Towers gives the visitor an intimate, friendly impression," so unlike the detached awe he felt when visiting the Biltmore Estate, George W. Vanderbilt's massive mansion in western North Carolina; McArdle felt that Gray should "capitalize on the possibilities of enhancing this friendly visitor feeling for Grey Towers."[21]

To do so, however, required a substantial budget and securing those plentiful funds was going to be a struggle. McArdle wrote that he believed "in aiming high and think you have done this, but in the present fiscal crisis there is always the possibility of getting involved in an 'exercise in futility.'" On the bright side, as long "as your local congressman [Rep. McDade] is on the House Appropriations Committee," McArdle's "guess is that Grey Towers will not be cut off entirely. Even so for several years you are not likely to get funds for 'optimum' development." He cautioned Gray to think within bounds: "Building maintenance, not major or even minor construction, but just preserving what you have, will always have to get first priority," and then he would have to "decide whether or not to close the place to all visitors," though assuming "you get any money at all, I think you will have to provide for visitors."[22]

Protecting those core activities meant being pragmatic about blue-sky desires: the stated ambition to purchase additional property from the Pinchots, even if they had been willing, was a nonstarter: "money for land acquisition is going to be virtually non-existent in the future." Constructing a new administrative facility, visitor center, and conference site—all part of the optimum package—were also quite unlikely to secure the requisite funds despite Rep. McDade's keen interest in these projects. "This is only a personal opinion, of course, and may not be worth much. My suggestions are compromises from what I would like to recommend and a program you may have to retreat to," concluded McArdle, the thirty-year veteran of the Forest Service.[23]

McArdle's instinctive urging of restraint was born of his decadelong service as chief. He intuited that Chief Max Peterson (1979–1987) would act as he would have—and his intuition was spot on. During its first four years, the Reagan administration made a show of clamping down on some forms of federal spending; social services were cut, defense was not. The overall economy was also in a wobbly state, a surfeit of timber hit the market and prices plummeted, housing markets fell off sharply, and a serious bank crisis loomed. No matter how compelling the Grey Towers master plan was, no matter that Chief Peterson had a genuine affection for the site (for which he had been the chief engineer in charge of its initial renovations in 1963), the provisions of the 1982 master plan were not going to be first in line for agency funds.[24]

Edward Vandermillen would learn this firsthand when he succeeded John Gray as director of Grey Towers in 1983. He came to Milford after serving as head of Forest Management for the Northeastern Area of State and Private Forestry. Because State and Private Forestry, like the other branches in the Forest Service, was then in a period of retrenchment, Tom Schenarts, director at the Northeastern Area, shuffled some key staff to reconfigure its budget. With Gray's retirement, he shifted Vandermillen, whom he knew to be "interested in history, old books, and antiques," to head up Grey Towers.[25]

Although the site's budgetary cupboard was relatively bare, Vandermillen could bank on the social capital that Gray had amassed at Grey Towers, having recruited a cadre of museum curators and interpretative specialists from inside and outside the agency. Since 1978, Leo (Bob) Neville, Carol Drescher, John Denne, Amy Snyder, and Carol Severance had begun to implement many of the recommendations of the preceding years' strategic reports, particularly the historic preservation mission that the 1979 *Historic Structure Report* had identified as critical to the protection and exhibition of the Pinchot family's history and its connections to American environmentalism. These key individuals made Vandermillen's job, amid fiscal pressures, much easier.[26]

He made the site's work a bit easier in turn with a concerted effort to reach out to the Pinchot family. Crucial in this regard was the personal attention he paid to resolving Ruth Pinchot's ongoing concerns about the impact of visitors, whose numbers were rising. Vandermillen took seriously too Gifford Bryce Pinchot's long-standing lament that in the previous twenty years the site had never reached its potential as a catalyst for conservation, a think tank for new ideas designed to resolve the many looming environmental problems facing the country and the world. The spike in the number of regional and national conferences that occurred during Gray's tenure, many of which were associated with the Pinchot Consortium, continued under Vandermillen, extending the site's reputation as a neutral arena that was hard to locate in the 1980s, as highly polarized national debates erupted over such hot-button environmental issues as the Love Canal and Three-Mile Island disasters, acid rain and endangered species, and clear-cutting and wilderness preservation.

Then there was the effort to establish the Pinchot Foundation, a concept that emerged in 1977 as a result of the Forest Service deciding that a stand-alone nonprofit organization might develop into a collaborative venture that would fund-raise in support of a revived Pinchot Institute. In this endeavor it was seeking to construct a more sustainable relationship than it had developed with the Conservation Foundation in 1963, even though the goals were roughly the same—working with an effective partner to generate moneys for scholarships, fellowships, environmental education curricula, and the publication of various materials. As Grey Towers directors, Gray and Vandermillen were charged with fleshing out this concept, and in 1983 the Friends of Grey Towers was charted as a 501(c)(3) organization. Its tax-exempt status was critical to its philanthropic mission, and its political neutrality—also required by the federal tax code—dovetailed with the Forest Service's long-standing belief that the Pinchot Institute's work must be science-based and impartial. Although this iteration of a supportive network failed to materialize, this attempt, when combined with the painstaking interior and exterior maintenance projects, makes it clear why Gifford Bryce Pinchot believed Grey Towers had never been so active.[27]

It explains too why he was so startled in 1986 to learn that the Forest Service Chief Max Peterson had tapped Vandermillen to head up the agency's Office of Public Affairs: "The Pinchot family and I are deeply distressed that Ed Vandermillen has been ordered to leave the Pinchot Institute. He is the one person who has made the Institute into a going concern." Among Vandermillen's achievements, Pinchot wrote, was the development of "joint operations with Yale's School of Environmental Studies"—whose dean, John Gordon, had utilized the landmark for conferences and retreats. He praised Vandermillen for turning the fledgling Friends of Grey Towers into "a significant force during his tenure" and for the "real start" he made in "restoring some of the aesthetic and historical values that were destroyed under previous managements." Grateful for Vandermillen's nurturing of Grey Towers as an intellectual hub, Pinchot also had a more personal reason for being aggrieved: "not least of his achievements has been to restore a really friendly and warm relationship with the Pinchot family and to restore their belief in the ability of the Forest Service to do an imaginative, educated, and intelligent job of using Grey Towers." Reiterating that "previous managements did not enjoy this friendship and faith," he suggested that Vandermillen's departure might spark a controversy the embattled agency could ill afford: "At a time when the Forest Service is being attacked so heavily by environmental groups, the reversion to mediocre management at Grey Towers can only give the Forest Service another, but nonetheless legitimate, reason for being attacked."[28]

The abilities that so attracted Pinchot to Vandermillen were among those that Peterson wanted on his staff. But the chief may have had another reason for bringing Vandermillen to the Washington Office. By so doing he was opening the way to promote a woman, Rita Thompson, to be Grey Tower's director, a strategy that shaped how Thomas Schenarts, Northeastern Area director, would fill this job. Once an almost all white-male enclave, in the early 1970s the Forest Service slowly, grudgingly began to hire women outside the classic steno pool—as rangers who fought fires, cut trails, or pushed paper. "As late as 1976," the historian James G. Lewis observes, "women held eighty-four percent of clerical jobs in the agency and fifteen percent of administrative and technical jobs, but fewer than two percent of full-time professional jobs." This low glass ceiling did not get its first crack until 1978, when Geraldine Larson was appointed deputy forest supervisor at the Tahoe National Forest; the decision was so unprecedented that the regional forester who made the final decision first sought assurances from Larson's husband that her promotion would not disrupt his San Francisco–based business. Seven years later, Larsen became the first female forest supervisor. So tilted against women and minorities was the agency's culture, and thus the opportunities for promotion, that in 1973 a group of women in Region 5 (California) filed a sexual discrimination lawsuit in federal court. After six years of negotiation, the agency accepted a consent decree that required it to increase the number of women in all levels of its workforce to equal that of California itself, which then stood at 45 percent. This judicial victory was Pyrrhic: the blowback within the agency and rearguard actions of the Reagan administration stymied the upward mobility of women and stirred up a poisonous atmosphere that did not dissipate until the late 1990s. The 2001 appointment of Sally Collins as the agency's associate chief (its second in command) and the 2007 appointment of Gail

Kimbell as its sixteenth chief—the pair of women held these top offices until 2009—marked the final breakthrough at the Forest Service.[29]

Grey Towers was not immune to these tensions. Vandermillen, a long-time agency employee, was at times inept in his working with the site's staff, many of whom were women. Well trained and eager to expand their responsibilities and advance their careers, they found him less than even-handed in his treatment of their professional insights and aspirations. Observed one male colleague: "Ed was pretty old school in some things, and he did not realize how some of his statements and behaviors were being taken by some of the women." While there is no direct evidence that employee concern about Vandermillen's management style expedited his transfer to Washington, Chief Peterson's decision to replace him in 1987 with Rita Thompson, making her Grey Towers' first female director, is noteworthy; her presence there diversified this particular administrative post. It is hardly coincidental that this personnel decision came the same year that the agency released a report on its future employment objectives titled *Work Force 1995: Strength through Diversity*. Yet it weakened its claim to this moral high ground when it elevated Thompson, then serving as a district ranger on White Mountain National Forest—one of only five in the agency's entire eastern region—by cutting her new position's pay scale. John Gray and Ed Vandermillen had been paid at a GS-15 level; Thompson began at a GS-13 level. She was not the first woman in the Forest Service—or in wider society—to learn that equal work did not earn equal pay.[30]

Questions of gender at Grey Towers and within the Forest Service also were bound up with an equally striking difference in the way that its female and male employees articulated the range of their environmental commitments. The distinctions were revealed in an intriguing set of attitudinal surveys conducted from the 1970s to the 1990s. What they revealed was that women who held traditional jobs in the agency (e.g., forestry), like those who were in nontraditional occupations (such as the curators and landscape architects at Grey Towers), were much more likely than their male peers to advocate for a reduction in timber production on the national forests, an increase in protection for old-growth forests, and an expansion of the number of acres receiving wilderness protection. The concept of landscape stewardship, as reflected in these women exhibiting a "greater environmental concern than men," added another layer of stress to the workplace, whether in the field, lab, or office.[31]

The analogy to Grey Towers was the pivotal role that women there played in recovering, promoting, and preserving the site's historical texture and biographical fabric. Indeed, they wove together these two threads, recognizing that their twinned relationship was manifest most of all in the lives and activism of the women of the Pinchot family. Carol Severance was among those who contributed to this expansive perspective.

She joined the Grey Towers staff in 1981 as part of the team Gray was assembling to restore the house; she credits her immediate supervisor, Bob Neville, with shielding her from most of the ramifications of the old boy network that then dominated the Forest Service. Severance and her colleagues pushed back in their own way: their investigations into the house's architectural history and evolving material culture led to a forceful rewriting of the male-dominated narrative surrounding Grey Towers that the agency previously had promoted. Instead of simply being the mansion that James Pinchot paid for, Richard Morris Hunt designed, and that Gifford Pinchot later inhabited, Severance, along with her colleagues Amy Snyder and Carol Drescher and the consulting archivist Jean Pablo, demonstrated that Mary Eno Pinchot and Cornelia Bryce Pinchot actually had been central to the mansion's conception, construction, and regeneration, inside and out. To build this case, Severance sifted through the Pinchot family's papers in the Library of Congress and in the family's private collections, and she presented her findings to the docents, staff, and many of the conference groups who met at Grey Towers. In so doing, she was helping to fulfill the site's larger mission of developing and transmitting new scholarship to enrich visitors' educational experience.[32]

In reclaiming Cornelia Pinchot's feminist principles, conservation ethics, and political activism, Severance also found her voice—which she enhanced during a two-year leave of absence so that she could write her master's thesis on James Pinchot's art collection. One of her compelling arguments is that James's patronage of Hudson River School artists—Sanford Gifford, Eastman Johnson, and Worthington Whittredge, among others—and his admiration of their vision of nature as a restorative balm, led him to embrace the conservationist perspective he and his future wife would pass on to their children (a transmission signaled when they named their first born after Sanford Gifford). In recovering this link between James Pinchot's artistic sensibilities and his environmental convictions, and in her archival reconstitution of the artworks that once hung at Grey Towers, Severance also subtly critiqued the Forest Service's crude handling and hasty disposal of the collection when it took over Grey Towers in the early 1960s;

its ill-advised actions robbed the site of a priceless cultural resource and invaluable teaching tool.[33]

For all these successes, the physical site continued to deteriorate. Asbestos ceilings that the Forest Service had slapped up in the early 1960s had to be removed, the slate roof leaked, the rough-stone foundation seeped, and the Bait Box and the Letter Box—two outbuildings that Cornelia and Gifford had built—were in disrepair; rehabbing the expansive grounds and gardens had been put on hold. Short on staff and budget, the staff's immediate to-do list lengthened.

Mirroring this disarray was the low morale of those who worked at Grey Towers. Its energy-sapping effects emerged during Vandermillen's tenure as director, and they grew under Thompson's leadership, continuing when Thompson retired in 1988 and Vandermillen returned for a brief stint. It did not help matters that several key figures either had been promoted to the Northeastern Area headquarters or had left for graduate school, or that Thompson alienated the Pinchot family. By the late 1980s, the situation had become so unhealthy that Forest Service leadership arranged for a site evaluation by Leroy Johnson, a field representative for the agency's Northeastern Area. He confirmed Grey Towers' dysfunction. While he was escorted around the mansion and grounds, Johnson recalled, "I was not introduced to another member of the staff, and although my guide could tell me the names and titles of the other staff, he couldn't (or wouldn't) tell what their main responsibilities were. I left with an uneasy feeling and concluded that Grey Towers was a 'White Elephant.'"[34]

What a difference a few years—and new leadership—made. Returning to Grey Towers in 1992 for yet another evaluation, Johnson was struck by what he encountered. "Probably nowhere within the Forest Service could you find a more enthusiastic and dedicated staff," he wrote in his second review of the site's operations. "Today, every person on the staff is friendly and outgoing to visitors. Everyone could tell me what Grey Towers' mission is . . . and they know the goal of their program area. They also knew what other staff responsibilities were because they had worked with others on various teams." Its revived "esprit de corps" was the direct result of the intense "training and leadership" that the new director Edgar Brannon had introduced when he arrived at Grey Towers two years earlier. He had "inherited a fragmented staff, who, by their own admission, did not work together as a team to accomplish the Institute's mission or their respective program goals." With each having "their own turf to protect," there was "scant cooperation across program lines." Breaking down those barriers was essential if Grey Towers was to function as its advocates long had hoped.

Brannon had come to the job not certain what to expect. A landscape architect, and the first one to become a forest supervisor, he had managed more than five hundred employees at the Flathead National Forest in Montana, and he did so with an outsized budget. There, Brannon was daily enmeshed in some of the most intense conflicts anywhere in the national forest system—logging companies and environmentalists routinely brawled with one another and the Forest Service over timber-harvest levels, endangered-species protections, recreational opportunities, energy production, wildfire management, and wilderness preservation. The *New York Times,* among other publications, featured these struggles: "At stake, say the proponents of development, are thousands of jobs and the well-being of hundreds of small towns across the West whose workers rely on the timber and minerals of the national forest system." Environmentalists countered that "unless their perspective is heard the vistas that many of them moved west to enjoy may be threatened, species like the grizzly bear and the wolf may be diminished, and a tourist industry that may one day replace today's struggling sawmills as a source of jobs will never grow." Feeding into these tensions, and exacerbating them, was the Reagan administration's natural resource agenda, which was to accelerate cutting, drilling, and mining on public lands. Caught between those who he called "utilitarians" and "naturists," Brannon likened that contemporary debate to the one in the early twentieth century that ultimately had divided Gifford Pinchot and John Muir, founder of the Sierra Club. "This is not something that was first invented since Earth Day."[35]

That Brannon liked being in the center of action is clear and that is why he was wary about Grey Towers being a backwater. Still, the volatility and stress of his job at Flathead took their toll such that when old friend and former Grey Tower staffer Bob Neville invited him to serve on a strategic-planning committee for the site, which included Peter Pinchot and Nancy Pittman, grandchildren of Gifford and Amos Pinchot respectively, he agreed. It was a short, two-week detail, but as had happened to his predecessors Brannon come away entranced: "Grey Towers seemed like a semi-academic job and also such a peaceful place." He noticed too that its downsides had an upside: "A completely dysfunctional organization waiting to be saved. Right up my alley." A year later, when Northeastern Area director Michael Rains offered Brannon the directorship, he accepted it, a transition that came with more than a little buyer's remorse; after serving on a national forest over which everyone fought, he had accepted a job at a place that was but an afterthought.[36]

To put it at the forefront required a radical shift in workplace culture and managerial ethos, a subject that in the 1980s and 1990s had become a part of the national discourse. "Interest in management theory went into overdrive," Adrian Wooldridge argues, because the "rich world in general, and the United States in particular, tried to come to terms with the rise of Japan, the spread of computers, and the invention of new financial techniques." Because his ideas had done so much to transform postwar Japanese corporate culture, Edward Deming's principles of Total Quality Management (TQM), which stressed that all personnel, top to bottom, should be engaged in the process of improving the delivery of services, were copied throughout American businesses. Gaining adherents too were John Kotter, the author of *Leading Change,* which underscored the dynamic interactions necessary to bring about social transformation, and Edgar Schein, at MIT's Sloan School of Management, who weighed in on how managers could identify, evaluate, and then rebuild the constitutive elements of their enterprises. Joining this vigorous dialogue were Peter Drucker, Tom Peters, and, intriguingly, Gifford Pinchot III; for them all, entrepreneurial innovation was essential to organizational rebirth. Beset as it was by a range of critical forces internal and external, in January 1992 the Forest Service committed itself to pursuing TQM across the agency (and in time also would hire Gifford Pinchot III and his wife and collaborator Libba Pinchot to help with its Clinton-era reinvention process).[37]

That same month Leroy Johnson conducted his evaluation of Grey Towers, made reference to the agency's new managerial commitment, and anointed Grey Towers its poster child: "Empowered, creative, innovative, customer driven, and managing for total quality are precepts that characterize the working philosophy of the Institute's staff." It had reached this point because shortly after his hiring, Brannon—a devotee of Schein's and Kotter's reorganization strategies—immersed his staff in the prevailing concepts of team-based management. As they assimilated these concepts over the next year, he and his new colleagues developed a working model that included new job descriptions and flattened hierarchies, incentivized innovations in the site's environmental education offerings and conference attendees' experience, and fostered interagency cooperation in the reclamation of the site's historic built and natural landscapes. "We can disagree without fear of repercussions and we have no fear of expressing our opinions. We do what we think is right, not what we think the Director wants," a longtime employee told Johnson. One recent hire gushed to Johnson: "Coming to Grey Towers from a national forest was like coming out of the dark. The team approach works—it's great!" For once, Grey Towers was ahead of the curve.[38]

The same could not be said of the building itself, but its then-dilapidated state would only serve as a baseline for the remarkable effort the Grey Towers staff undertook to bring it back to life. They had plenty of evidence of what not to do. The 1979 *Historic Structure Report,* for example, had set up a definitive restoration objective: "as much as possible, the property should be returned to its pre-1963 appearance," to the time of "Pinchot ownership." It fudged on how the site would achieve this glorious end when it urged the creation of a "long-range construction program" that depended on "routine maintenance . . . to help accomplish these restoration objectives." Instead of a full-fledged strategic plan that laid out how, step by step, dollar by dollar, complete restoration would be achieved, it made do with an ad hoc process: "as various parts of the buildings, such as roofs and walls, require repair or replacement, the work should be carried out in a way that conforms with the overall goals of the restoration on the property." Even the simple standards this approach promoted required considerable capital investment and careful management. Yet despite spending "vast amounts of construction and maintenance money," the infrastructure continued to crumble. Walkways were buckled, exterior woodwork was rotting, and significant leaks remained. "Well intended, but improperly executed, rehabilitation measures have been carried out in the past," Johnson confirmed. When the agency contracted out the "seemingly simple job of repointing the chinks on the exterior wall of the Letterbox building, [it] exacerbated the plaster failure on an interior wall." The reason for this failure was that neither the contractor nor the Forest Service recognized that if "repointing was done with modern mortar that is waterproof" it would create "a moisture barrier, which caused the interior wall to collect moisture." Not understanding that complex interaction pointed to a larger problem: "This ever-increasing backlog of rehabilitation projects is now seriously degrading the spiritual image of Grey Towers," conveying—falsely, Johnson asserted—that the "Forest Service does not maintain this historical landmark."[39]

That was the deplorable situation when Brannon arrived. Over the next decade he and his team would completely rehabilitate Grey Towers, a consequence of intense planning, the securing of major funding from the Forest Service and an extraordinary set of congressional appropriations, and cross-agency collaboration. The first need was a complete evaluation of the fabric of the mansion and its outbuildings, from attic to basement, roof to ground; included in this analysis was a detailed report on the condition of the entire 101-acre property—woodlands and gardens, patios, pathways, roadbeds, and a local cemetery in which the first two generations of Pinchots were interred. Guiding Brannon on this process was William Klein, the director of the Morris Arboretum in Philadelphia and the chair of the board of the National Friends of Grey Towers. In the mid-1970s Klein had been hired to revitalize the arboretum, and over the next decade and a half he had reframed the site's mission and raised funding and public support for what had been a neglected resource for the University of Pennsylvania. Regenerating the physical landscape and associated historic structures, Klein also built a robust set of educational programs. Modeling Grey Tower's rehabilitation after Klein's successful efforts, Brannon and his colleagues launched an intense, yearlong in-house conversation about the site's central objectives, the first of which was held at the Morris Arboretum and out of which emerged three key goals:

— To steward Grey Towers much as ecosystem management guided foresters to ensure the ongoing integrity of the ecosystem, or in this case the buildings and grounds

— To interpret the life and legacy of Gifford Pinchot

— To partner with the Pinchot Institute to carry that legacy forward

With that framework in place, the Forest Service announced a national design competition for the production of fully fledged plan for the restoration of Grey Towers so that it could fulfill its mission. The winning proposal was a collaborative bid by a pair of Philadelphia-based firms, Andropogon Associates, an ecological planning and design firm whose task was to develop strategies "for the adaptive reuse and preservation of the historic landscape," and the Vitetta Group, which specialized in historic preservation, planning, and design. Beginning in 1993, the two firms identified how to recover the historic furnishings and infrastructure and what kinds of exhibit space, conference facilities, and staff offices could be constructed without compromising the structure. Their joint Master Facilities Site Plan provided a clear pathway for a staged process by which to reclaim the Pinchot estate. At $250,000, the Forest Service got its money's worth.[40]

The agency was also cued as to how much the overall project was going to cost: a staggering $18 million. To make the case for full funding, and to seek an emergency infusion of moneys to redo the dilapidated slate roof so as to protect Grey Tower's interior spaces and public safety, Brannon met twice with the Forest Service's chief and staff during August and September 1994. The first meeting was a rude awakening. As he walked the leadership through the landmark's staggering array of needs, and ended with the projected price tag, he was met with an ominous silence—and then a cacophony of voices explaining why his requests, large and small, were impractical and impossible. The cost was unimaginable to an organization that secured annually $20 million for all its facilities' maintenance (and half of that was dedicated to salaries). As the senior leadership beat up the proposal and beat down its implication—standard procedure in this highly competitive environment in which deputy chiefs fight to protect their respective turfs and outlays—Chief Jack Ward Thomas tracked the debate. He had had a long relationship with Grey Towers, starting with his association with the Pinchot Consortium. As head of the Amherst, Massachusetts, research unit in the mid-1970s he had had some oversight of the Pinchot estate and its administrative assistant who managed day-to-day operations, and frequently he had visited it while attending conferences and workshops. An adherent of Gifford Pinchot's practical conservationism, Thomas routinely quoted the first chief of the Forest Service in public speeches and internal memoranda, arguing that to succeed in the modern era the agency must recapture the "Pinchot thrust of leadership." It is not surprising, then, that he was sympathetic to Brannon's claims. Breaking into the fray, he suggested that Brannon return with a streamlined proposal that addressed some of the concerns that chief and staff had expressed. The meeting adjourned without resolution.[41]

Two weeks later, Brannon was back for what he expected to be a make-or-break conference, and his approach to it reflected its high-stakes nature. He challenged the group with this paradox: the agency was happy to affirm that Grey Towers "was an asset, but treated it as a liability." Having never capitalized on its asset value, it had never been able to shake its liabilities. But whatever choice leadership made—to keep and rehab or let it go (and several in the room had proposed getting rid of Grey Towers)—the cost would remain the same. Either the Forest Service would find ways to pay for the rehabilitation of the site or if the National Park Service took it over, as would have been the most likely scenario, attaching it to the nearby Delaware River Water Gap National Recreation Area, that agency would demand that the Forest Service shoulder the costs out of its budget. The double threat of having to admit that it refused to protect Gifford Pinchot's home and seeing its founder's manse, which President Kennedy had dedicated to renewed public service in the cause of conservation, moving over to the Park Service shut down the critics. When Brannon left the room asking for emergency funds to rebuild the leaking roof, Thomas, a Texan native who, like Lyndon Johnson, was partial to language colorful and eliminative, declared he was "not about to be the chief who pissed on Gifford Pinchot's grave."[42]

Although Brannon received almost a million dollars to repair the roof, left unresolved was his request for the millions needed to rebuild Grey Towers. For that funding, the agency's initial strategy had been to seek the support of Congressman Sid Yates (D-IL), a longtime member of the House Appropriations Committee, chair of its Interior subcommittee, and a good friend of the Forest Service. The November 1994 elections, in which the Newt Gingrich–led GOP gained control of the House of Representatives, disrupted this possibility; Yates was now the ranking member of the relevant committee. Plan B was to approach Rep. Joe McDade, now the senior Republican on Appropriations, and seek a steady stream of what is called "no year money." These funds rolled over from year to year, granting Grey Towers maximum flexibility for how and when it spent federal dollars. What Brannon hoped to avoid was for Grey Towers to receive earmarked appropriations, which would have resulted in moving funds from one Forest Service account to another, generating ill will among his colleagues. McDade made the full restoration possible, Brannon later observed: "he kept Grey Towers from going under—we ought to name it after him!"[43]

As these federal funds, along with grants from Pennsylvania's Department of Natural Resources and private philanthropies, flowed into Grey Towers over the next seven years, Brannon hired National Park Service historic restoration crews to rebuild the roof and other vital portions of the buildings' fabric. The foundation and basement were shored up, new HVAC systems were installed, windows were reglazed, and wall-sized murals that the Forest Service had painted over in the 1960s were carefully recovered. As National Park Service curators helped Grey Towers staff identify and purchase furnishings consistent with what archival photographs and familial correspondence indicated once had graced the mansion, Brannon gave local antiques dealers a standing order to purchase any furniture, artwork, or objet that had been sold or pilfered from Grey Towers in the early 1960s (with some success). The grounds and landscaping were reworked just as extensively. Landscape architects unearthed old garden beds and replanted them according to Cornelia Bryce Pinchot's designs; cuttings were taken from aged apple trees, Park Service horticulturalists grafted them onto new root stock, and they were then replanted around the estate. A new entrance roadway, paths, and trails were constructed to lead visitors up the hill from a well-screened visitor parking lot and informational pavilion, and guests received much more comprehensive interpretative programming about the site and its transformation.[44]

The Pinchot family took notice. They hardly could do otherwise—their onetime home landscape was in the throes of a multimillion-dollar upgrade. But they were wowed and wooed in others ways. In June 1990, following Gifford Bryce Pinchot's death the previous August, Grey Towers staff developed a memorial tribute "to express our appreciation to the family for his gift of Grey Towers." The exhibit, which was placed in the Bait Box, the small building his parents had built for their son, displayed documents, artworks, welded objects, home movies, books, and photographs that his widow, Sally Pinchot, lent for the occasion. The curators' ambition was to evoke his life at Grey Towers, his professional career and intense love affair with oceanic sailing, and his environmental activism. Following the family's planting of a memorial apple tree in Pinchot's honor, Brannon spoke briefly about the forthcoming restoration projects, then in the conception phase, and his commitment to "carry out the programs Dr. Pinchot had in mind when he donated Grey Towers to the Forest Service." Building on this initial outreach, Brannon drafted brothers Gifford III and Peter and their cousin, Nancy Pittman—and in time other members of the rising generation—to participate in the site's ongoing planning processes. In 1993 Peter addressed the thirtieth anniversary celebrations for the Pinchot Institute and spoke movingly of his family's deep roots in this land; in collaboration with the Forest Service, subsequently he restarted the Milford Experimental Forest that his great-grandfather James had established as part of the Yale School of Forestry summer school on the property. Reflective of the reintegration of the family and the agency was the official commitment in the site's Strategic Plan from 2000 to 2005 to sustain this close relationship.

This was the first time that any of its many planning documents had made nurturing this connection an explicit goal. The crescendo came on August 11, 2001, on what would have been Gifford Pinchot's 136th birthday. After a round of speeches applauding the completion of the complex rehabilitation project, Edgar Brannon swung open Grey Towers' original nine-panel, three-hinged, double-wide wooden door, welcoming the assembled throng, including four generations of the Pinchot family, into the sunlit foyer. At that moment the emotional response Peter Pinchot had expressed at the 1993 rededication was even more poignantly apt: "I wish my father and mother were alive to see this."[45]

CHAPTER TEN

Neutral Force

Issues of democratic health are intimately connected to ecological sustainability.

—Hannah J. Cortner

When the Grey Towers staff sat down to map out the relationship of
the national historic landmark to the Pinchot Institute, they captured
it with a Venn diagram. Named for John Venn, who in 1880 had refined
the concept of an overlapping set of circles to represent the connections between two or more seemingly distinct sets of ideas, objects, or, in this case,
institutions, the tool helps illustrate what is logical and probable at any particular moment. In 1999 it made sense for Grey Towers' staff to highlight
the lines of connectivity between the federal agency and an organization
it helped bring to life in 1963 but which had since then undergone a series
of transformations: the onetime conservation education initiative had become the Pinchot Consortium in the early 1970s and a decade later was
reimagined once again through the creation of the National Friends of Grey
Towers; by the late 1990s the Pinchot Institute was no longer a unit of the
Forest Service located solely in Milford, but it had morphed into a stand-
alone 501(c)(3) organization with offices at Grey Towers and in Washington, D.C. To facilitate its work, the organization received an annual grant
from the federal agency to underwrite policy analysis it conducted at the
Forest Service's behest and pursued grants from a variety of governmental and private entities to work on such issues as third-party certification
of forest-management practices and ecosystem management. To illustrate
this change in the institute's status that resulted in a new relationship with

the federal agency, the Venn diagram, embedded in Grey Towers' Strategic Plan of 2000–2005, revealed the principle points of contact and integration—the overlay. Among these were policy analysis, leadership training, meeting facilitation, and the convening of environmentally and natural resource themed conferences and symposia. Many of these concepts were the same ones that the Forest Service, and its then-partner the Conservation Foundation, had expected to develop at Grey Towers in the 1960s; they also had formed the basis for every one of the agency's subsequent attempts to fulfill the statutory obligations and commitments it took on when it received the Pinchot family's gift of the Milford estate. The Venn diagram was an attempt, in short, to illustrate this long-standing link over time.[1]

Yet Venn diagrams by definition are also static; they lift identifiable relationships out of time when they set them down in a bidimensional space. So too with the Grey Towers model: this clean graphic masks the historically messy and baffling relationship between the Forest Service and the Pinchot Institute, a tangle of politics, personality, and power, of budgetary shortfalls, creative leaps of imagination, and organizational shortcomings, of luck good and bad. It also does not explain why, seemingly against all odds, the institute somehow survived until, and was even thriving at, the beginning of the twenty-first century.[2]

Yearning for that exact result, yet hardly in a position to predict such an outcome, were the men and women who gathered in the Hart Senate Building on Capitol Hill in late February 1984 to attend the first board meeting of the National Friends of Grey Towers. Its leadership had met two months earlier in Milford to develop the organization's bylaws and governing structure, where they learned from Forest Service Chief Max Peterson that the Reagan administration had zeroed out Grey Towers' budget, a stark deficit the agency hoped this friends' group could erase through fundraising to sustain the site's conferences, restoration projects, and educational programming. Board member Peter Pinchot affirmed his family's rekindled interest in their "historic home and the community," and its eagerness to help the new group return Grey Towers to its onetime status as a place where "eclectic individuals would gather to debate unique ideas and issues." They had heard as well from Rep. Joseph M. McDade (R-PA), who reiterated his decadeslong commitment to Grey Towers: "Each year I have worked to set aside additional funds to assist basic maintenance and visitor services. But these funds are not adequate. With visitor rates ever increasing, additional financial support beyond the federal share is required—not only to restore and maintain the home itself but to expand its educational programs." Recognizing the added bang that would come if it held its public launch in the nation's capital, board member Harry Buchanan, vice president of the Celanese Corporation, urged Rep. McDade and the national chairman of the group, Pennsylvania's Lieutenant Governor William W. Scranton III, to invite the entire Pennsylvania congressional delegation and a select number of other well-placed representatives and senators to attend. It is not known who had the brilliant idea to schedule a pre-lunch cocktail hour with these influential guests, but surely it lubricated the conversation and perhaps smoothed the way for ensuing financial commitments to the fledgling organization.[3]

More sobering news awaited: the National Friends of Grey Towers needed every penny it could scrounge, reported John Barber, its inaugural president and retired deputy chief for State and Private Forestry and currently executive vice president of the Society of American Foresters; it had received donations totaling $7,213.60. That amount could not begin to support the organization's very ambitious plans to fund a cooperative relationship with the Yale School of Forestry and Environmental Studies, including scholarships for students who would spend a year at Grey Towers to assist its operations, beef up of the site's environmental educational programming for all ages, and nurture Grey Towers as a much-needed "neutral ground" where leading policy experts could meet "to discuss and resolve worldwide issues." With that paltry sum, the board could neither underwrite the creation of a traveling exhibit of James Pinchot's remaining art collection nor support full-scale restoration of Grey Towers' material culture and landscape aesthetic. To top off this list of as-yet unfunded commitments, the Friends of Grey Towers proposed establishing a publication fund and an endowment, for a total budgetary goal in its first year of $220,150. The fundraisers never came close to closing the gap between what it had in the bank in February 1984 and what it hoped its balance would be in December—or any other subsequent year.[4]

Part of the problem was that the severity of the Reagan budget cuts, which had been the impetus for the group's establishment, was wreaking havoc with the National Friends' ability to secure major donations from philanthropists and foundations. The president had argued that private donors would step in and replace the loss of federal dollars for cultural institutions like Grey Towers. The opposite proved true. Knowing that Washington could not be bothered to support a federally owned historic landmark, potential contributors shied away from funding what was perceived as a losing proposition. Harry Buchanan had warned Chief Peterson of this catch-22 and had pushed the agency to maintain the site's budget at $500,000, believing this would send an important signal to well-heeled givers, individuals and corporations alike. Peterson could only come up with $295,000 (all but $30,000 of which was dedicated to salaries and benefits), admitting that this was minimal support for minimal programming. "Harry, I know this is not the level of commitment you had hoped for," wrote Thomas Schenarts, director of the Northeast Area, "but it does demonstrate that the Forest Service has a commitment to Grey Towers even during these difficult times when many programs are being eliminated." Not surprisingly, potential contributors interpreted partial funding as partial, complicating Buchanan's already difficult pitch for donations.[5]

Unable to bring in the necessary outside investment, the National Friends turned to Grey Towers' best friend in Congress, Rep. McDade. By law, the Forest Service could not spend more money than was appropriated for its operations, and it could not directly request additional dollars. The National Friends were under no such constraint, however, and became an essential conduit between the Forest Service and the representative's office. McDade staffer Debbie Weatherly, for example, posed a question through Sidney Krawitz of the National Friends' finance committee that he relayed to Grey Tower's director Edward Vandermillen: "What is the distribution of the FY 1984 funding for Grey Towers and what are the priority maintenance projects for major health and safety items?" Vandermillen responded with a three-page, itemized budget that set down what the landmark was currently able to spend, and he went on to identify more than one million dollars of deferred maintenance to the house, outbuildings, grounds, parking lots, and roadways that if unfixed would seriously compromise the buildings' integrity and visitors' and staff's safety. To stay within the law, the groups' executive secretary forwarded this data to Weatherly with the all-important opening sentence: "As you requested from Ed Vandermillen, the enclosed information shows categories for the current budget . . . and a prioritized list of maintenance needs."[6]

Krawitz meanwhile was drafting language that McDade could employ to make certain that the Forest Service's budget for Grey Towers reflected the reality of its many needs. "It is the suggestion of the National Friends of Grey Towers that Congressman McDade requests the following direction be given relative to 1985 funding for Grey Towers," and it was a suggestion that came with specifics: "Funding for Grey Towers and the Pinchot Institute for Conservation Studies will be at the same level in Fiscal Year 1985 as in Fiscal Year 1984, and will be funded through benefitting functions. In addition, a minimum of $100,000 will be used for maintenance of health and safety from the budget line item for maintenance facilities. Furthermore, maintenance should be added each year in addition to basic operations budget presently at the $500,000 level." The language's precision was deliberate and necessary: to finesse the Reagan administration's budget cutters, the Forest Service employed the National Friends to request Congressman McDade to direct the federal agency to boost its budgetary allocation for Grey Towers. This pattern continued until the budgetary pressures eased in the late 1980s, with McDade using his position on the House Appropriations Committee to influence the Forest Service's budgetary decision making, each year successfully directing it to make certain that Grey Towers had $500,000 for operations and maintenance.[7]

To secure those funds meant something had to be cut: one of the programs that the Forest Service began to strip money from was the Pinchot Consortium, which received its last federal dollars in 1986; scientific research's loss was Grey Towers' gain. This did not mean the site was flush—far from it. With more than half its budget committed to salaries, Grey Towers staff had to make do with little. That is why the National Friends' fundraising efforts were so crucial, and why its flagging efforts were so frustrating. Still, the historic interiors were slowly preserved, one item or detail at a time. The historic landscaping was similarly on a slow track to recovery. Amy Snyder, who wrote a detailed, ten-year plan for the step-by-step restoration of Cornelia Bryce Pinchot's gardens, had a mere $9,000 a year with which to work, making every penny count as she and volunteers patiently reconstructed the central elements of "surprise, anticipation, and illusion" that Cornelia had infused into the estate's gardens, walkways, meadows, slopes, and patios. Meanwhile, the National Friends secured a significant annual gift from a funder to create a "living history interpretative program," money that helped bring audio-visual specialist and actor Gary Hines to Milford temporarily to refine his award-winning, one-man show *Pinchot;* he became a member of the permanent staff in the early 1990s and took his show on the road, performing before hundreds of audiences across the country, an educational outreach of inestimable value.[8]

The conservation studies aspect of Grey Towers' mission lagged, understandably so; its 1984 budget had no money encumbered for research, conferences, or fellowships, and succeeding years were as bereft. Once more, outside funders intervened and their intervention set into motion the first steps that would revitalize the Pinchot Institute and lead ultimately to its emergence as an independent policy institute. With support from its onetime collaborator the Conservation Foundation and funding from the Howard Heinz Foundation and the Forest Service, the Pinchot Institute for Conservation Studies at Grey Towers (its official title on and off since 1963) hosted the first of what it projected would be an annual series of national conferences. Using as its template the multinational conversations between high-level officials in the United States, Canada, and Great Britain held each year at Ditchley House in England, the Grey Towers confabs were expected to draw "leaders in government, industry, academia, and conservation organizations [to] discuss issues related to the long-term productivity of the nation's natural resources."[9]

The initial meeting, held in mid-September 1986, focused on population change, natural resources, and regionalism, a topic framed around an older set of worries that Conservation Foundation founder Fairfield Osborn had probed in his 1948 best seller *Our Plundered Planet* and that Paul and Anne Ehrlich had revised and updated in *The Population Bomb* (1968); one scholar has argued that the Ehrlichs' provocative tract was responsible for "climaxing and in a sense terminating the debate of the 1950s and 1960s" that Osborn had launched. Although the debate's intensity may have diminished by the mid-1980s, consumption of natural resources—wood, oil, coal, water, minerals, and grass—had only accelerated. Figuring out how to calculate the impact of this pressure, when combined with swift population migration, and how to respond to these paired forces, was the conference's aim.[10]

At least that is what the conferees argued, and they placed their arguments within a historical framework and geographic context. After forty years of sustained dispersal of population, leading to a steady decline in density in the urban core and an expanding and thickening periphery, rural environments and economies were reeling. Top-down federal management of such quandaries in the 1960s, as reflected in the Kennedy-Johnson era in which the Appalachian Regional Commission and the New England River Basins Commission were created, appeared less persuasive twenty years later with the anti-federalist thrust of the Reagan administration. Seeming to promise greater hope was the concept of state-driven regionalism, as exemplified in initiatives to resolve acid rain and a nongovernmental approach in which nonprofits and/or business interests would collaborate across state lines. "I'm not a fan of formal regionalism," Arthur M. Davis of Penn State noted, and so he promoted instead what he called "ad hoc regionalism," episodic cooperation and coordination of specific issues, among them "water resource planning; waste disposal; air quality . . . and other economic and environmental issues of opportunities of common concern."[11]

Another proponent of a more bottom-up strategy was R. Neil Sampson of the American Forestry Association. "It has become apparent during the course of our discussions that the nature of governance needs to be clarified," an issue that also was dominating the national political discourse. "Decentralization has been furthered by the Reagan administration," Sampson asserted, "but it was not invented by it." Whatever its source— and he noted that the many organizations represented at the conference as well had been shifting their onetime national foci to the local level—it no longer was clear how this change in orientation would impact Washington's leadership. Perhaps the Forest Service, with its decentralized structure, could serve as a model: its virtues "should be brought out in the open to help other agencies." In this context, the past held no sway, Sampson concluded. "Certainly we are all stirred by Gifford Pinchot's legacy, but we must remember that Pinchot's day is long gone, as is the time when we can emulate his methods. His was a day when elites provided leadership," and the challenge in the mid-1980s was whether and how national leadership "can be exercised in an era of decentralized operations and actions."[12]

A provocative disclaimer, Sampson apparently was unaware that as governor of Pennsylvania in the 1920s Pinchot had engineered the formation of a five-state Ohio River compact to manage water pollution along the river's full extent, an innovative solution preceding by decades the federal government's water-regulation legislation. It seems Sampson was unaware as well that Pinchot had been responsible for the very decentralized structure of the Forest Service that he had so lauded; Sampson's "future trend," in which "Washington-based offices . . . function in a support role for their field offices," depended on the first chief forester's top-down management decision in 1905 to do just that. History was not yet past.[13]

That irony aside, the 1986 Grey Towers conference proved influential, internally and externally. It gave a boost to the underfunded Grey Towers staff that recognized they could successfully convene such an event; the National Friends were similarly pleased that it was able to pull together the funding for and gain a higher profile from the conference. The gathering also introduced a new generation of natural resource analysts to Grey Towers, notably those from the Conservation Foundation. Among them was the foundation's new president William K. Reilly, who had worked under Russell Train on the Council of Environmental Quality during the Nixon administration and one year earlier had merged the foundation into the World Wildlife Fund (which Train was running); in 1988 he would follow Train's lead again, becoming administrator of the Environmental Protection Agency. Key staff presenting at the conference included Robert G. Healy, who specialized in land-use and environmental regulations and shortly thereafter would direct Duke University's Center for Resource and Environmental Policy, and William E. Shands, who with Healy had coauthored *The Lands Nobody Wanted,* a seminal study of the eastern national forests. Shands later would join the Pinchot Institute as a senior policy fellow and in time encourage one of his younger colleagues at the Conservation Foundation, V. Alaric ("Al") Sample, to become its president. The community of conservation policy analysts was closely knit.[14]

The National Friends continued to struggle, this tight weave notwithstanding. Securing needed dollars proved difficult, compromising its ability to meet its stated goals and objectives. There were organizational challenges too: it had a president and board of directors that set policy and pursued funding opportunities, but there was no dedicated staff to carry out these initiatives. Its precise relationship with the Pinchot Institute was puzzling too and was reflected best in the official stationery of the two entities: the header for one read National Friends of Grey Towers, with the Pinchot Institute for Conservation Studies as its footer; the other letterhead flipped this top-bottom relationship on its head. Outsiders, and even some insiders, must have been confused by this lack of clarity, which only complicated the two groups' combined operations and logistics.[15]

This fog began to lift in 1989 as a result of two personnel decisions. The National Friends board of directors, conscious that it could accomplish little as a volunteer board without any staff, held a make-or-break meeting with Forest Service Chief Dale Robertson. The question was simple: Should the National Friends and thus the institute be disbanded or would the agency be willing to provide annual funding to hire the organization's first executive director and sustain the office? Robertson, a supporter of Grey Towers, the Pinchot Institute, and by extension the friends group, agreed to provide a modest annual grant for administrative support to be leveraged by additional private fund-raising, and the National Friends board hired James W.

Giltmier to guide its operations. Giltmier's hiring coincided with a change at Grey Towers: Edgar Brannon, forest supervisor on the Flathead National Forest, became its new director. Together, these two individuals would alter the integrated institutions.[16]

A former journalist, Giltmier was a longtime staffer on the U.S. Senate Committee on Agriculture, Nutrition and Forestry, which had brought him into close contact with the Forest Service. He had also served as a legislative assistant for Senator John Melcher (D-MT) and had been a Washington representative for the Tennessee Valley Authority. As a reporter and back-office operative, Giltmier was well trained in the give-and-take of American politics and schooled too in the need not to take credit for his efforts: when he wished to express an opinion in the *Conservation Legacy*, a newsletter he initiated under the National Friends' banner, or in policy statements and routine correspondence, he often did so through the guise of a "close friend" or a stand-in figure he dubbed "The Troll."

However framed or voiced, Giltmier's credentials and contacts would be invaluable to furthering the National Friends' advocacy of conservation policy in the nation's capital. To make this happen, Giltmier transformed the composition of the organization's board. Policy analysts and academic and resource specialists were brought aboard, replacing those who had originally been most interested in the restoration of Grey Towers. To reinforce this reorientation, Giltmier created the Pinchot Council, an informal advisory group comprising some of the rising stars in forest policy in the United States—Berkeley's Sally Fairfax, Syracuse's Margaret Shannon, and Hannah Cortner at Arizona, along with key foresters such as Lolo National Forest supervisor Orville Daniels, Lassen National Forest supervisor Kent Connaughton, and Al Sample. Marking these changes was one of nomenclature—the National Friends of Grey Towers stopped using that name and began to call itself the Pinchot Institute for Conservation.[17]

This new identity created a dilemma because Edgar Brannon's official Forest Service title was director of the Pinchot Institute of Conservation Studies at Grey Towers. As a work-around, Brannon agreed to use the title director of Grey Towers, while Giltmier would be known as the institute's director. The split would become official in 1995, with the formal emergence of the Pinchot Institute for Conservation as an independent, tax-exempt, nonprofit corporation. That year marked another transition: Al Sample took over the reins of the Pinchot Institute; Giltmier stayed on as a senior fellow. With Sample, the position became professional.

A forest economist, with advanced degrees from the Yale School of Forestry and Environmental Studies and Yale's School of Organization and Management, Sample had worked in the industrial sector for Champion International, in government for the Forest Service, and as a policy analyst at the Conservation Foundation, which he left following its absorption into and later extinction by the World Wildlife Foundation. Before coming to the Pinchot Institute, he had been vice president for Research at the American Forestry Association, the nation's oldest forest-conservation advocacy group (established in 1875, one of its early officers was James W. Pinchot, Gifford's father). Closely connected to the D.C.-based policy world, Sample built off these connections to lead the Pinchot Institute into several critical debates shaping forest policy in the mid-1990s.

The most contentious revolved around the basis by which the Forest Service would manage the 193 million acres of national forests and grasslands under its management. The explosive controversy over clear-cutting, which Gifford Bryce Pinchot had helped fuel in the early 1970s when he challenged the agency's harvesting techniques in his father's name and which led to a seminal lawsuit, *Izaak Walton League v. Butz* (1973), continued well into the 1990s. It was supposed to have been resolved by the passage of the National Forest Management Act of 1976 (P.L. 94–588)—segments of which James Giltmier had worked on—through its promotion of extensive public involvement in forest-planning processes and its requirement that the Forest Service conduct full inventories of lands that then were zoned by use; it also required, among many other steps, that the agency identify alternative strategies for land management, including a category that the Forest Service, by its name and can-do culture, has had a hard time embracing—"no action."[18]

Although the agency and many citizens expected the new forest-planning processes to calm the waters, the opposite occurred. "Typically, plans stimulate controversy," wrote Shands, Sample, and Purdue University researcher Dennis LeMaster in a 1990 report for the Forest Service, "because the process surfaces new information and generates proposals that benefit some interests and threaten others." If the scales tilted against them, grassroots environmental groups and national organizations like the Sierra Club and the Wilderness Society used the act's oversight provisions to challenge individual forest plans in the federal courts. The Reagan administration exacerbated this situation by pushing the agency to accelerate timber harvests, to get out the cut regardless of the environmental consequences (in the careful language of then-chief Dale Robertson: the GOP "was not too keen on making policy changes, especially if it would adversely affect economic benefits from the national forests"). Adding to the furor was the emergence of an internal dissident group, Forest Service Employees for Environmental Ethics (FSEEE), scientists and planners who disputed official findings and actions and leaked evidence of flawed research and illegal decisions that imperiled endangered species and the protection of biodiversity. The war in the woods seemed entrenched.[19]

The legal logjams were symptomatic of the agency's larger confusion about its managerial mission and social purpose. Hoping to chart a less contentious path, Chief Robertson adopted a concept his staff called "New Perspectives." It depended on the idea of ecological forestry that the forest ecologist Jerry Franklin, a longtime agency researcher in the Pacific Northwest, and Tom Crow, in the Great Lakes region, had been implementing on the ground. They argued that the Forest Service needed to view the acres under its stewardship as ecosystems set within a broader landscape and manage these varied terrains to perpetuate the survival of their many and interconnected elements. Healthy forests included a wide range of tree ages and types, downed logs and snags, open areas and thick canopies, and leafy riparian buffers. Harvesting should reproduce these beneficial ingredients, practices Robertson confessed he did not fully appreciate after inspecting some of Franklin's initial work: "It was certainly different and in many ways looked like a sloppy, unfinished logging job!" Of course, that was Franklin's point—the desire for a slick and clean harvest made a mess of nature.[20]

Robertson announced his conversion in congressional testimony in 1989, promising that this more ecological approach to land management would become standard across the national forest system. The next year, the Pinchot Institute convened at the agency's request a conference on New Perspectives out of which came an official "charter" that encouraged land managers to sustain ecosystems "for a wider variety of benefits and uses now and in the future" and affirmed the Forest Service's commitment to open up the planning process even more than it had done in response to the National Forest Management Act, a vow that was coupled with its assertion that "all aspects of natural resources conservation are to be integrated" into land-use planning. With this permission to act, district rangers, forest supervisors, and other line officers started testing Franklin's and Crow's ideas on their landscapes, with more than 260 such projects under way by 1992.[21]

Although Robertson recollected that this was a fairly seamless transition, in truth there was an intense pushback against his decision inside the agency and within the executive branch. The "decentralized structure of the Forest Service," observes the historian James G. Lewis, "the lack of funding for new projects on many national forests, and the failure to clearly define New Perspectives at the beginning made implementation challenging."[22]

Its external critics heightened the tension, blasting the idea as greenwashing, a clever ploy to hoodwink the public, a rebuttal they reiterated several years later during the 1992 United Nations Conference on Environment and Development, held in Rio de Janeiro, Brazil. In the first days of the Earth Summit, EPA administrator William Reilly, who headed the U.S. delegation, was pounded by U.S. and global environmental activists, who attacked what they believed was the Forest Service's mismanagement of the nation's public lands; how could the United States critique poorer countries' environmental records when its own actions were so inimical to healthy ecosystems? Through administrative channels, Reilly reached out to the agency, seeking policy clarification; Robertson responded that the Forest Service, under most circumstances, "was willing to eliminate clear-cutting as a standard method of timber harvest on the national forests." Several days later in his speech to the Rio conference, President George H. W. Bush affirmed that this proposal would become national policy.[23]

What exactly did ecosystem management mean? How and where did it converge, if it did, with the similar management concepts embodied in New Perspectives? To untangle these and other issues, the agency turned to the Pinchot Institute to conduct a survey of its actions, including face-to-face interviews and focus groups with land managers about their perceptions of New Perspectives and ecosystem-management principles it endorsed. In retrospect, some of its findings were not earth shattering. The majority of respondents felt that these policy changes emanated from outside pressure (e.g., the emergence of a more aggressive environmental movement), and that the biggest obstacles to the agency's operations were lodged in the federal budget and congressional-set production targets. As one interviewee observed, "unless expectations for outputs change, even strong verbal mandates will have little effect." Yet those in support of ecosystem management suggested they saw the inherent benefits that could come from its expansion as policy. A strong cohort agreed that New Perspectives allowed more "social values" into management decision making, a healthy shift in agency philosophy. Those on the ground—some of whose projects pre-dated Robertson's articulation of the new modus operandi—also viewed the changes positively; a "convenient vehicle of action," ecosystem management "pushed us to look at some things we wouldn't have otherwise." Concluded the study's authors: "In encompassing human social systems, New Perspectives—and Ecosystem Management—shrugged off the constraints of old definitions of resource management in order to maintain relevance in an increasingly complex and demanding society."

Their collective reactions confirmed to a degree the subsequent arguments of Hannah Cortner from the University of Arizona, who served on the technical advisory committee for the Pinchot Institute's analysis of New Perspectives. Arguing that ecosystem management is "dramatically different from traditional, sustained-yield resource management, with its focus on utilitarian values," it approaches nature with "some reverence and respect for the awesome complexity with which its components are interwoven." In so doing, its first priority is the "protection of ecosystem attributes and functions," and the conserving and sustaining of these values across time; significantly, "levels of commodity and amenity outputs are adjusted to meet that goal." The Pinchot Institute's James Giltmier concurred with Cortner's perspective: "these ideas are a refinement of Gifford Pinchot's ideals of conservation, an exploration into how our species ought to relate to nature."[24]

Where Cortner parted with Giltmier and the Pinchot Institute report's authors highlighted a key difficulty in their analysis. "If Forest Service officials are unable to carry through with the promise of New Perspectives—or, perhaps, worse, if Ecosystem Management degenerates into a standard set of technical practices backed up by page after page in the Forest Service manual," Cortner argued, then this bureaucratic inertia will disillusion its proponents. The cost will be multigenerational, for the agency "will suffer a setback to creativity and innovation that will be difficult to overcome." More difficult, Cortner believed, were the outside structural impediments that could hobble the promising, even radical program. Because it subordinated commodity production, ecosystem management was unlikely to secure congressional favor and robust appropriations; because it appeared to challenge the way that large timber companies and livestock operations made use of the national forests and grasslands, it was unlikely to secure their favor. States and communities where these public lands were located would also have to be convinced that the loss of receipts flowing off the federal forests was a good thing. This led her to determine that the wide-scale implementation of ecosystem management "will require extensive social and political changes" and would have to include a "redefinition of the values that define relationships among humans and nature, professions and citizens, and government and citizens." Trickier still was the pursuit of this complex set of changes at the same time as the reforming or "dismantling of traditional resource management institutions, such as agencies and laws." To fulfill its potential, New Perspectives and ecosystem management would require a paradigmatic change in American culture.

New paradigms do not emerge overnight. For proof, note that ecosystem management, as a concept and management technique, has remained controversial decades after its initial articulation. Yet in that time managerial strategies on public and private forests have shifted. One signpost of this change has been the increased demand for third-party certification of what the Pinchot Institute calls "conservation-minded forest management." A direct response to consumer pressure for sustainably harvested wood products, itself an outgrowth of the national debate over ecosystem management, governmental agencies and woodland owners sought tools by which to evaluate and confirm that their practices qualified for a "green" imprint. To meet the standard, argued the Pinchot Institute's Sample and Roger Sedjo, a forest-policy analyst at Resources for the Future, producers had to satisfy three conditions. Their techniques simultaneously must be "ecologically sound, economically viable, and socially responsible." Such an integration was critical to successful ecosystem management and reflected the lessons that Gifford Pinchot had learned at the beginning of the twentieth century and that environmentalists absorbed at its close—that it was not possible to maintain "long-term protection of forest ecosystems without incorporating the economic and social needs of the local people into conservation strategies." To realize this ambition required a set of criteria and checklists that were more or less rigorous in their calculation of how carefully landowners applied the triple bottom line to their forestry practices. In the United States the two prevailing strategies have been the independent Forest Stewardship Council (FSC) and the industry-supported Sustainable Forestry Initiative (SFI).

The Pinchot Institute's contribution to this development began in 1996 when it "embarked on a long-term research project to see whether certification programs—originally developed to guide forest management and timber harvesting by private companies—could also help improve forest management on public lands designated to protect a wider array of natural resource and environmental values." Its first contract was with the Pennsylvania state forest system, for which it audited its entire 2.1 million acres. The process allowed the institute to tweak its assumptions about how certifications were conducted and how its results would be interpreted, and the Keystone State, once it instituted the changes that the Pinchot Institute had recommended, became the world's largest single owner of certified forest—more than 3,000 square miles. The success of that project led to another, far more massive one. Over the next ten years, the institute provided certification audits of five units of the Forest Service on which it tested the applicability of the FSC and SFI models; North American tribes hired it to evaluate their sustainable practices on 30 separate units of land stretching across the country; the institute worked with the state of Oregon to determine the feasibility of establishing a formal "Oregon Certification Standard" for the Beaver State's 28 million forested acres, public, tribal, and private; and the state of Maine brought it in to assess whether its forest service could facilitate the certification of small, family-owned forests. These contracts, and others like them, allowed the Pinchot Institute to contribute significantly to the necessarily incremental effort to build consensus for, by demonstrating the value of, ecosystem management.[25]

As indicative of the organization's maturation was the development of a more solid and constructive relationship with Grey Towers. Much of this revolved around the friendship that emerged as Sample and Brannon worked together, coming to a shared conviction that the two institutions functioned best when their operations were in close dialogue. Part of that conversation, as it had earlier with the National Friends of Grey Towers, turned on the Pinchot Institute's ability to serve as a go-between for Grey Towers on Capitol Hill, in the Forest Service, and more broadly within the executive branch. That function came in handy after Brannon analyzed the historic up-and-down relationship between the Milford site and the agency and concluded that Grey Towers' budget was too dependent upon the whims of Congress and varying levels of goodwill in the agency leadership. Hoping to stabilize what had been mercurial, attempting as well to institutionalize the site's cooperative relationship with the Pinchot Institute, in 2002 Brannon and Sample met with James Snow, special counsel for Real Property in the Office of General Counsel for the U.S. Department of Agriculture, seeking a legislative fix.

The goal was to draft legislation that would enhance Grey Towers' legal standing, perhaps as a national monument or some other designation that would, Snow remembered, "give the property more status in the budget battles and require the agency to make Grey Towers a full-fledged component of the National Forest System." Its then-current notation as a National Historic Landmark, which it had gained in 1966, did not offer the kind of protection and identity that Brannon was seeking. Snow determined that the most suitable legal term would be to call it a National Historic Site. There was a catch: the National Park Service would object that only those properties in its inventory could secure that designation. It had said the same thing about national monuments, but its claim in that regard had been upended with the Alaska National Interests Lands Conservation Act (1980); that law created the Admiralty Islands National Monument and Misty Fiords National Monument under Forest Service administration. Snow decided that this might set a precedent for National Historic Sites, and accordingly titled the proposed legislation the Grey Towers National Historic Site Act. Introduced in Congress in 2004 by Rep. Donald L. Sherwood (R-PA)—Joseph McDade's successor—it came with the full support of the Pennsylvania delegation. During markup in the House Committee on Natural Resources, the National Park Service objected, but the committee sustained Snow's judgment about the designation's suitability. Because the act did not come before the 108th congressional session ended, Rep. Sherwood, a member of the House Appropriations Committee, appended the bill to the Consolidated Appropriations Act of 2005. With its passage, Grey Towers became the first National Historic Site outside the Park Service's purview and thus the first within the Forest Service's domain.[26]

More salient were some of the act's critical provisions. It allowed the Forest Service to acquire on a willing-seller basis any acreage, adjacent or otherwise, from the Pinchot family, should it in the future be interested in doing so. This potential expansion was matched with an unusual budgetary finesse: "Subject to such terms and conditions as the Secretary [of Agriculture] may prescribe, any public or private agency, organization, institution, or individual may solicit, accept, and administer private gifts of money and real or personal property for the benefit of, or in connection with, the activities and services at the Historic Site"; this meant that these funds were for the sole benefit of Grey Towers, "a notable exception to the Miscellaneous Receipts Act," Snow noted, which requires that "all funds received by a federal agency be deposited as general receipts in the U.S. Treasury." Finally, the act explicitly confirmed and affirmed a tight relationship between Grey Towers National Historic Site and the Pinchot Institute for Conservation: "The Secretary is authorized to enter into agreements for grants, contracts, and cooperative agreements as appropriate with the Pinchot Institute, public and other private agencies, organizations, institutions, and individuals to provide for the development, administration, maintenance, or restoration of land, facilities, or Forest Service programs at Grey Towers or to otherwise further the purposes of this Act." Grey Towers' director Edgar Brannon (and his successors) had been given the authorities he had hoped to secure.[27]

But then Brannon had been laying the groundwork for this possibility for several years. The first step of which had been that very Venn diagram gracing Grey Tower's strategic plan for 2000 through 2005—which Brannon had prepared and which his direct supervisor, Michael Rains, director of the Northeast Area, had recommended and Chief Michael Dombeck had approved. Their signatures on this document's cover page legitimized this diagram's implied aspirations; six years later, the Grey Towers National Historic Site Act gave them the force of law.[28]

CHAPTER ELEVEN

Common Cause

The challenge, and it will continue to be a challenge, is how do we keep
things in balance. How do we support a community, how do we keep an
industry alive, and then how do we do that in a sustainable fashion so that
at the end of the day your ecosystem remains intact?

—Catherine Mater

It comes down to the land, its health and viability, its capacity to regenerate and sustain its ecological relations and their integrity. If salubrious
and energetic, then the communities—biotic and human—depending on
them will flourish. If not, then the consequences could be destabilizing.

That was the message Gifford Pinchot's parents conveyed to him on
his twenty-first birthday when they presented him with a copy of George
Perkins Marsh's *Man and Nature,* a clarion call for an informed conservation stewardship that James and Mary Pinchot promised to enact on the
many acres surrounding their just-opened Grey Towers estate. Aldo Leopold made a similar claim in *A Sand County Almanac* (1949) about the pressing
need for a self-conscious ethic, the best definition of which, he suspected,
was "written not with a pen, but with an axe. It is a matter of what a man
thinks about while chopping, or deciding what to chop. A conservationist
is one who is humbly aware that with each stroke he is writing his signa-
ture on the face of his land." This sensitive engagement required as well a
sense of one's small place within the long sweep of time, an insight Leopold voiced in 1907 while participating in the Yale Forest School summer
camp at Grey Towers; he was happy to "pick up the axe again," he wrote
his mother, and "while I am biting into the heart of a big pine or chestnut,
to think that each chip is like a chip cut out of the interval between Now
and Then." Some of those intervals are more pivotal than others, President

John F. Kennedy asserted in 1963 as he accepted Grey Towers on behalf of a grateful nation; there are eras that especially require the presence of purposeful actors, and the early 1960s appeared to be one of them: "For our industrial economy and urbanization are pressing against the limits of our most fundamental needs: pure water to drink, fresh air to breathe, open space to enjoy, and abundant sources of energy to release man from menial toil."[1]

Yet determining how to act in good conscience necessarily has always been a work in progress, an evolution of human understanding of what constitutes healthy ecosystems. So reasoned the small gathering of "farmers and philosophers, foresters and theologians" meeting at Grey Towers in 1991. They had come together because they "wanted to look over the edge, into the distance to see if we could come up with some common ideas about future stewardship of the land." Out of their concerted efforts emerged the Grey Towers Protocols, a set of principles that were designed to speak to what its authors believed was the third stage in the "history of Americans' relationship to their lands and natural resources." The first—the getting and taking—had led to the violent subordination of native peoples and the equally devastating and rapid clearing away of forests and other resources to fuel what in time became the agricultural and industrial revolutions of the nineteenth century. The second had been born in response to the first: as habitats disappeared, hunters and anglers fought to protect woodland, meadow, and marsh; as large mammals and avian life were exterminated, in their defense women and men banded together to form the Audubon Society and similar grassroots organizations; and local needs for clean water or flood protection or fire control led still others to advocate for creation of national forests, parks, and refuges. One key result of this agitation was the passage of the Creative Act of 1891, which marked the creation of the nation's first forest reserves (later national forests). This law's centennial, and a reappraisal of its significance, had brought the conferees to Grey Towers to hammer out an updated approach to resource management for a new century.[2]

Participants at the two-day conference recognized that the late twentieth century liquidation of tropical rainforests and the "plight of endangered species" had compelled their generation to "recognize that forest ecosystems—and the life forms that inhabit them—have values that transcend those of direct human utility." This observation became the lynchpin of the four main planks of the Grey Towers Protocols:[3]

1. Land Stewardship must be more than good "scientific management"; it must be a moral imperative.

2. Management activities must be within the physical and biological capabilities of the land, based on comprehensive, up-to-date resource information and a thorough scientific understanding of the ecosystem's functioning and response.

3. The intent of management, as well as monitoring and reporting, should be making progress toward desired future resource conditions, not on achieving specific near-term resource output targets.

4. Stewardship means passing the land and resources, including intact, functioning forest ecosystems, to the next generation in better condition than they were found.

The Pinchot Institute adopted these principles as its guiding philosophy and they have framed its efforts ever since. This embrace has not been by happenstance—V. Alaric Sample, then on the staff of the Conservation Foundation and later to become the Institute's executive director, served as the conference amanuensis and authored its summative work *Land Stewardship in the Next Era of Conservation*.

The new focus on ecosystems as the baseline for management came with a significant challenge: Would these principles work on the ground and in real time? Demonstrating their applicability has been a considerable undertaking and has led to an array of projects to test their effectiveness and whether, if successful, these management concepts could be transplanted to other regions and societies. Ecosystem management, in short, may be a scientific notation, but its success depends as much on site-specific biota as it does on place-based human concerns that are social, economic, and cultural in origin and articulation. This meld added a political dimension and aspirational component to the work to come. Noted Sample: "As important as these principles might be in guiding the physical activities of resource managers on the land, they might be even more valuable as a means for resource managers to communicate a vision of stewardship and personal responsibility to society at large, helping a fragmented public to recognize that our economic well-being as well as our environmental health rests on our being able to pull together rather than pull apart."[4]

Trying to stitch together the American polity has been a difficult, onerous, and not always fruitful operation, but that did not and does not mean the effort is misguided. Just that it has been and will remain incomplete.

More measurable and perhaps more fulfilling has been the work itself, as suggested in four case studies of the Pinchot Institute's activities in rural forested watersheds in North and South America. Each has played to and off of the ethos embedded in the Grey Tower Protocols. Each has found its source in a commitment to a local, land-based community need. Each has attempted to develop a broad coalition in its support. Often operating under the radar and on private lands, a strategy that has allowed their managers, researchers, and supporters to develop their project objectives outside the oft-contentious national debates between economic development (jobs) and environmental protections (wilderness), they have been able to experiment with solutions, devise correctives, and expand their potential influence. As these initiatives have been transplanted from their sites of origin they are helping to make less central those furious arguments that since the late 1960s frequently have defined the contentious parameters of American environmental culture.

That a Pinchot is involved in two of these projects and that these initiatives make explicit reference to earlier generations' efforts to regenerate battered lands offer a compelling story line. It is just not one that Peter Pinchot thought would be his. "For the first thirty years of my life, I resisted a Pinchot career in forestry," he confessed in a 1999 speech to the National Leadership Team of the U.S. Forest Service; "it seemed far too daunting to be in my grandfather's shadow. But eventually the green conservation blood got the best of me." He completed a master's in forestry at the Yale School of Forestry and Environmental Studies in 1997, which his progenitors had established. When, two years later, he officially joined the Society of American Forests, an institution his grandfather Gifford Pinchot had founded, he laughed: "my last defense was punctured. I am afraid I am a fallen man."[5]

What he fell into was the felt need to reconceive conservation from the ground up. His reconceptualization would take into account the late twentieth-century shift away from a commodity-based approach often associated with his grandfather's defining notion of the forester's creed: "the greatest good of the greatest number in the long run." It would make use too of Aldo Leopold's land ethic that depended on an alteration in the "role of *Homo sapiens* from conqueror of the land-community to plain member of it." Taking seriously the fundamental change in the relationship between Americans and nature was an essential precondition to rethinking how the Forest Service might better steward the national forests and grasslands. The majority of Americans by the mid-1960s, Pinchot argued, were "living in urban and suburban areas and had little daily dependence on nature for their livelihood," and thus they had "no direct exposure to the raw products of forests"; for them, the "idea of managing forests for a sustainable flow of commodities no longer had much personal meaning." These urbanites found more resonant the new environmental ethos pitting "environmental protection and wilderness preservation against the economic thrust of natural resource productivity."[6]

Pinchot was not the first to draw this close connection between social change and ideological recalibration, and he was not unique in arguing that this dynamic had put the Forest Service on the defensive. The resolution he proposed to the agency's leadership—that it use the 193 million acres under its stewardship to protect the planet's diminishing biodiversity and at the same time provide ways for city residents "to reconnect to the wildness of real natural landscapes"—was also in line with then-chief Michael Dombeck's commitments. One of the chief's goals, dubbed "forest to faucet," had been developed to enhance urbanites' realization of how deeply connected they were to forested watersheds that supplied much of their potable water (in the American West, for instance, more than 30 percent of water supplies flow off national forests and grasslands). "We can leave no greater gift for our children," Dombeck asserted, or "show no greater respect for our forefathers, than to leave [the] watersheds entrusted to our care healthier, more diverse and more productive." What Peter Pinchot brought to the conversation about how the Forest Service might regain its onetime relevance and centrality was to argue for a communitarian approach to landscape management.[7]

This new paradigm called on foresters "to reexamine some of our core assumptions." To reduce the pressure on forests, "while we develop the scientific knowledge of how to preserve biodiversity in working landscapes," Pinchot urged his professional peers to push the wood-products industry to develop sustainable substitutes for "virgin wood fiber for reading materials, house construction, and packaging." Coupled with that charge was a more radical argument. The "model of multiple-use may have outlived its utility," he affirmed, especially in the context in which logging would devastate already identified "biodiversity hotspots, where a majority of the world's species are found." Pinchot proposed developing a zoned approach to timber management. Taking bio-rich areas completely out of production and transferring that work to locales "of low priority for biodiversity conservation," where it would be possible to "maximize fiber production to meet economic demands," made good sense. Recognizing too that private lands must also contribute to this recovery process, he suggested that it would be critical to develop "community-based programs that would provide incentives for neighboring landholders to restore and sustain species diversity on their lands."[8]

It was in this latter context that Pinchot would test his arguments. Teaming up with his family, and with financial support and technical advice from the Forest Service and the Pinchot Institute, he relaunched the Milford Experimental Forest that his great-grandfather James Pinchot had established in 1903 at Grey Towers to facilitate the field training of Yale forestry students. As reconceived in 1999, the site's mission focused on ecological restoration and community forestry. This orientation was a result of two central concerns that Pinchot and his collaborators hoped to address on the forest. Milford and the Pocono Mountains of northeastern Pennsylvania were quickly becoming a haven for second homes for those of means who lived in the New York metropolitan region, housing construction that was fragmenting older patterns of ownership and eroding some of the area's biodiversity. This demographic and biological challenge came coupled with a political one—how to engage residents old and new, year-round and seasonal, about the need to preserve the "esthetics and environmental quality of the region" that the growing tourist and recreational economy depended on and threatened. Given that "there are probably only one or two decades of opportunity to conserve enough of the landscape in blocks of continuous forestland to sustain the diversity and richness of the forest and river ecosystems," the Pinchot family placed a conservation easement on the bulk of the 1,400 acres it owned along the Sawkill River in which the experimental forest was located. Doing so allowed the family to demonstrate its commitment to sustaining the land's integrity and to "stimulate a regional dialogue about stewardship and to create a pattern of collaboration between private and public landowners that can begin to reverse the trend towards fragmentation of the forest."[9]

As transformative were the management strategies Peter Pinchot established for the site. Among the problems confronting the eastern woodlands was overabundant white-tail deer population, which in Milford as elsewhere had resulted in a clearing away of the forest understory. A two-year deer population study concluded it was necessary to launch a "large, managed hunt" to cull the herds on the property. Aware that the experimental forest could not solve the deer-density problem on its lands alone, Pinchot worked with scientists at Pennsylvania State University, the Pennsylvania Game Commission, and local hunting clubs "to develop a community-based deer management program with adjacent private and public landowners so that we can collectively manage deer at the landscape scale."[10]

Restoring the American chestnut locally has led to the creation of a similarly broad network of allies. The Forest Service and the Pinchot Institute, as well as the American Chestnut Foundation and the Connecticut Agriculture Research Station, have contributed time and expertise to the Pinchot family's effort to revitalize what had been the area's dominant tree species, whose nuts once lay thick on the ground. But sometime around 1900, the chestnut blight fungus (*Diaporthe parasitica*) was brought to North America and began to wipe out the tree, felling billions; it continues to plague efforts to return to the canopy a species once praised as the sequoia of the East. That moniker also has been a source of hope. Because the American chestnut "is a charismatic species," Peter Pinchot suggests, "the act of its restoration can help catalyze community stewardship of the regional forest."[11]

At the Milford Experimental Forest, like a number of other locations from New England to the Mississippi River Valley, the restoration effort has required a two-prong approach. First, the development of an Asian–American hybrid chestnut bred to be blight resistant, scientific experimentation that the American Chestnut Foundation and the Connecticut Agricultural Research Station conducted with some success. The other critical step has been to figure out how to reintroduce this species into the mature eastern forests from which the chestnut had been absent for a century or more. Earlier tests at Milford demonstrated how difficult this process would be. White-tail deer browsed on chestnut seedlings and sprouts and the blight continued to cut back the regenerative capacity of older trees. Subsequent efforts appear to be more a bit more successful, as fencing and hunting have keep white-tail deer populations under control. Harvesting of sunlight-blocking maples has opened up the canopy and a number of other silvicultural treatments are being assessed for their efficacy. Among those doing this vital assessment work is Leila Pinchot, Peter Pinchot's daughter, another forestry graduate from Yale. After completing her PhD at Tennessee in 2011, she was hired by the Forest Service and the Pinchot Institute to conduct an ongoing series of chestnut restoration experiments in the Milford forest. This acorn did not fall too far from the family tree.[12]

That tree was also transplanted abroad, courtesy first of David Smith, one of the volunteers who developed some of the initial forest restoration studies on the Milford Experimental Forest. When he left Pennsylvania as a Peace Corps volunteer, he was assigned to the northwestern Ecuadoran community of Cristobal Colon; nestled in the foothills of the Andes, it is situated in a wet tropical forest. Knowing that its economy then revolved around the rapid harvesting of timber to be converted into farmland, as Smith headed south he carried a copy of Gifford Pinchot's *Primer of Forestry* (1899). Among the text's central points was the enduring impact that "destructive lumbering" can have on a forest. It injures young growth, Pinchot wrote, "provokes and feeds fires," and can "annihilate the productive capacity of forest land for tens or scores of years to come." To counter this destructive process required the adoption of more conservative methods of forestry, the central purpose of which, in Pinchot's words, was "to draw from the forest, while protecting it, the best return of which it is capable of giving."[13]

The conditions Smith encountered on the ground in Cristobal Colon, Ecuador, seemed similar to those Gifford Pinchot had encountered in late nineteenth-century U.S. forests. Home to three hundred families that own more than one hundred thousand wooded acres within the Rio Verde Canandé watershed, Cristobal Colon is a poor town in good measure because its residents could not afford to sustainably manage their timber resources. Smith's economic analyses indicated that commercial agents were paying roughly $0.10 per foot for rough boards: "even when farmers cut as many trees as they can, their families still make considerably less than $1,000 per year selling their lumber wholesale." This exploitative situation was magnified with the clearing out of the forest, which left behind unstable soil on treeless slopes; subsequent erosion undercut the families' ability to supplement their income through agriculture. Hoping to restore the ecosystems and the community's economic viability, Smith reached out to the Pinchot family. Peter and his second wife Nancy, and the couple's blended family, began to spend significant time in Cristobal Colon helping knit together an international coalition of partners. Fundacíon Jutan Sacha, the largest nongovernmental conservation organization in Ecuador, the Pinchot Institute, the U.S. Peace Corps, and the U.S. Forest Service collaborated with this isolated rural community to "sustain forestlands in that region and spark economic development."[14]

Out of this collaboration emerged a fund-raising effort that purchased "appropriately scaled tools to allow the community to begin producing finished wood products," among them a portable band saw that increased the marketable yield from each tree cut (and thus decreased the number that needed to be harvested). In addition to requisite technical training, the cooperating institutions have provided education in marketing and business management, as well as wood-product development. Through the 2004 establishment of a communally owned corporation, EcoMadera Verde, Cristobal Colon's residents began to turn out hardwood flooring, molding, and furniture, all higher-end products generating more profit than sawn boards. In subsequent years the marketing strategy has shifted to creating balsa wood, and one of the products that Peter Pinchot has been particularly interested in is the manufacture and sale of wood blades for wind-energy turbines. "By creating many new community jobs and providing families with a market for sustainably produced timber," Pinchot wrote in language his grandfather would have recognized, "EcoMadera is creating an economic alternative to pervasive forest exploitation."[15]

Pinchot knew that this shift in fortune would not be easy, for Eco-Madera operates "in the midst of well-organized market forces that support forest exploitation." To address the difficulty in raising capital and achieving profitability, in 2008 community shareholders agreed to restructure Ecomadera Verde; two new companies emerged, one based in the United States known as EcoMadera Forest Conservation LLC and a subsidiary in Ecuador, Verdecanandé, SA. The former is an "American social venture company with the capacity to grow rapidly through access to capital." With its initial investment coming largely from Peter Pinchot, the day-to-day work is managed by Ecuadoran professionals and the organizations remain tightly linked to their community of origin; they are required to engage closely "with community leadership through continued local ownership." The ultimate goal also remains the same: "to help reverse the loss of forests in this region by providing the communities with the tools and skills to build a viable local economy based on sustainable management of their working forests."[16]

Recovering the viability of the Rio Verde Canandé's forested watershed has been bound up with the effort to sustain its residents' health. Much of that initial work revolved around developing a more sustainable economic market for its forest products. But rural poverty also has health care implications. For many in Cristobol Colon, "the forest serves as a kind of health insurance," such that when one of its three thousand residents "becomes sick or suffers an accident, the forest resources are harvested to pay for medical care." The implications of this led two volunteers at EcoMadera, Ariel Pinchot (another of Peter's children) and Julia Przedvorski, to conduct in collaboration with residents the first community-wide survey of its associated health care issues, facilities, and resources. The resultant data revealed widespread disease, from malaria and typhoid to dysentery. It also indicated that the community was suffering from malnutrition and high child mortality rates. Hindering the securing of better health care was a series of interlocking geographical, economic, and social barriers. The Rio Verde Canandé watershed had no health care workers or clinics, the nearest facility was eight miles distant, and that clinic itself was understaffed and without adequate resources.

To reverse this debilitating situation, in 2008 EcoMadera, with the support of the Ecuadoran Ministry of Health and the Pinchot Institute, launched a four-year project to build, furnish, and equip Cristobal Colon's first health care clinic and secure a full-time nurse and part-time physician. "Health is a basic human right and a goal onto itself," Ariel Pinchot has observed, "however good health is also vital from a systemic perspective, without which poverty alleviation and natural resource conservation cannot occur. Healthy families and healthy forests are intimately connected, and one cannot hope to achieve either without addressing health conditions and the degradation of natural resources concurrently."[17]

Ariel Pinchot's paternal great-grandparents made similar claims eighty years earlier. As part of their paired response to the industrial devastation of Pennsylvania's forest cover and the economic collapse of the Great Depression, Governor Gifford Pinchot and the Commonwealth's First Lady Cornelia Pinchot promoted what they called "human conservation." Simply put, there could be no economic recovery if the working and living conditions of the state's most impoverished residents were not enhanced, and there could be no sustained recovery if Penn's Woods were not regenerated. Social justice, economic development, and public health went hand in hand, they argued, an argument as true in Pennsylvania in the 1930s as it is in early twenty-first-century Ecuador.[18]

These intertwined principles are also being applied in Vernonia, Oregon, a small community of 2,300 residents inhabiting a narrow valley in the Coastal Range through which the flood-prone Nehalem River flows. Located in Columbia County in the northwest corner of the state, "a gritty little timber town that was once home to the largest electric sawmill in the world," its high school mascot tells the story of the community's historic work: the Mighty Loggers. Although surrounded, as Tom Hyde, a county commissioner, observes, "by 28 miles of forests in all directions," Vernonia confronts double-digit unemployment and a high level of poverty; resource rich, the community is decidedly cash poor. This imbalance is not unusual among other rural, mountainous, and forested communities scattered across the nation. It shares as well another dilemma facing these towns: a little over half of the forests in the United States are privately owned, the majority of these lands are the property of individuals or families, and nearly 50 percent of these owners are over sixty-two years old. It is this aging population that in Vernonia and elsewhere controls a significant portion of the country's forested ecosystems. Over the next two decades, some portion of these lands will be sold to sustain their owners' health and welfare. As in Ecuador, timber will serve as a form of health care insurance. Upon their owners' deaths, these woodlands may be sold to pay off debts and/or transferred to the next generation.[19]

The implications of this developing situation are the subject of Brett Butler's research. Butler, a social scientist working for the U.S. Forest Service, and his colleagues have written a series of technical reports that have identified ownership patterns across the country, surveyed current owners about their intentions for and felt needs from the wooded lands they own, and have assessed their future management plans and looming prospects for the sale, gift, or donation of these properties. Key to understanding this data, Butler has argued, is recognizing how family forest owners perceive their lands:

> Despite what some of us might have learned in forestry school, timber production is not the primary reason that families own land. . . . Rather, the most important reasons . . . are related to the aesthetics and privacy the land provides and its importance as part of their family legacy. "Aesthetics" is shorthand for the enjoyment owners get from many facets of the land—the trees, the wildlife, everything about it. Many owners have a primary or secondary residence on their land and greatly value the privacy and solitude their forests provide. "Legacy" is their ability to pass the land on to the next generation: many owners have inherited the land from their parents or other relatives and would like to do the same for future generations.

But whether they are able to hand off their legacy to their progeny, whether these legatees are interested in maintaining these woods as woods, is the crux of the dilemma. With an estimated six thousand forested acres a day being sold off in the United States, and the prospect of that number climbing amid what is predicted to be the largest intergenerational land transfer in American history, this shift in ownership could have a deleterious impact on the capacity of these woodlands to provide essential ecosystem services, including carbon sequestration and water quality, amid a changing climate.[20]

These linked and troubling issues serve as the foundation of the Forest Health-Human Health Initiative that the Pinchot Institute launched in Vernonia in 2010. Senior fellow Catherine Mater surveyed the parental and next generation's perceptions of how they expect to manage their familial legacies. Using a new health clinic in Vernonia as the pivot point, Mater interviewed upward of 25 percent of all family woodlands owners and their children within a twenty-mile radius. Her interviewees confirmed national studies indicating that health care, and its costs, were their number one worry. This was compounded in Vernonia, for 80 percent of the parents and their adult children interviewed did not have "long-term health insurance and no plan in place to address this health care need." Instead both cohorts expected to use timber resources to pay for emergency and/or sustained care.[21]

Those same trees targeted for harvest offer a different, more sustainable resolution. Oregon State University researchers inventoried the lands in question and discovered that "a majority of these forests are at an opportune time in their growth cycle from a carbon storage perspective, with more than 5,000 acres of these coastal Douglas fir dominant forests being comprised of trees 30 years old or younger." Asked if they would consider participating in a "carbon for health care program," in which the land owners would manage their woods to increase their carbon-storage potential and outside investors would purchase this increase in the form of carbon credits, an income stream that would be targeted for the owners' health care needs, the majority of the Vernonia survey expressed considerable interest. The Forest Health-Human Health Initiative was born.[22]

It is the first project in the nation to exchange forest carbon for direct payments to landowners and surrounding communities. With seed funding from the U.S. Department of Agriculture, Regence Blue Cross Blue Shield, and the Kelley Family Foundation, the Pinchot Institute has entered into a Memorandum of Understanding with the American Carbon Registry to serve as an official site for carbon credits in the pilot project and developed the ATreeM® debit card coded such that carbon-credit dollars can only be used for health care payments. As of early 2013 the institute was in the final stages of developing marketing portfolios for carbon investors from such health care sectors as hospital systems, insurance, and pharmaceuticals. "We believe investors will be willing to pay more for carbon credits," Mater has observed, "that are linked to quantifiable social benefits coming in the form of direct payments to health care accounts for landowners and rural communities." If so, then projects weaving together ecosystem services, public health, and rural sustainability could prove a dynamic combination transferable to other regions and conditions.[23]

A similarly complex connection between environmental health, water quality, and landscape restoration informs Common Waters, a project that has placed the Pinchot Institute at the heart of a critical challenge facing the sprawling 12,800 square-mile watershed of the Delaware River. As the institute's home ground, it is apt that its inaugural codirector Matthew Brennan years ago identified one component in the current initiative's name and objective. "We sit here at the Pinchot Institute in Milford in a ringside seat as a great water drama unfolds," he told a youth conservation training session at Grey Towers in 1966, for the "beautiful Delaware River which flows through the valley below us serves as the source of water for many towns and cities along its shores, including Philadelphia." But in the mid-1960s there were once-unfathomed pressures mounting on the watershed. "Several of the largest tributary streams above us are dammed to provide water for New York City. In years of drought, New York City holds its waters. The Delaware drops to a trickle," a diminished stream flow that had profound implications: "The lowered Delaware lets ocean tides come up the Delaware estuary dangerously close to the intake pipes which serve Philadelphia." Although that perilous situation had not yet occurred, because New York annually released enough to "keep the Delaware above the danger level for Philadelphia," Brennan posed the obvious question: "Would [it] if New Yorkers were short of water?" His point was not rhetorical. The "worst is yet to come," he predicted," for even with "normal rainfall in the future, the increased populations will make supplies inadequate." More complicating still was that it was not certain that those who so heavily depended on this vital watershed recognized the need to keep the Delaware running clear and high. "We can clean up our rivers, but the costs will be fantastic," he told the young conservationists in the audience: "Will we want to pay them? Do we really care?"[24]

Fifty years later those questions remained relevant because the "great forested landscapes of the Northeast and the critical watersheds they contain are facing death by a thousand cuts." New York and Philadelphia metro populations have continued to swell and, more importantly, to sprawl outward into such communities as Milford in Pike County. That is not the only reason that the local forests are been harvested and bulldozed. Private woodlands owners along the upper Delaware, like their peers in Columbia County, Oregon, are older, less well insured, and likely to cut portions of their timbered property to pay for critical health care costs or unexpected expenses. These sales and the resultant land parcelization are decreasing tree cover, adding to water-treatment costs downstream. Common Waters' goal is to reverse this process: "for every 10 percent increase in forest cover in the headwaters, water treatment costs are decreased by 20 percent." To secure this savings will require a realization on the part of those who own woods and those who need clean water of their shared interest in these paired resources.[25]

The Pinchot Institute has been serving in that interpretative role since 2007, when it began to coordinate more than forty state, county, and town agencies, regional authorities, nonprofits, land trusts, and foundations in support of the Common Waters Fund. The Common Waters idea emerged out an initial request of the Pinchot Institute to facilitate a meeting between the Delaware Water Gap National Recreational Area and adjacent property owners. The park's leadership realized that its management decisions depended on how its neighbors were managing their lands, but none of the various entities or individuals groups were talking to the other. The conversation began at Grey Towers that fall, and the assembled group decided that the Chicago Wilderness project offered the most persuasive template for the Delaware watershed initiative. Chicago Wilderness embraces a significant tristate area, covering portions of Wisconsin, Illinois, and Indiana and has built up a 260-group coalition, binding together urban and rural interests with a collective mission "to restore local nature and improve the quality of life for all living things, by protecting the lands and waters on which we all depend." Its actions testified as well to the capacity of voluntary partnerships to transcend political, social, and demographic boundaries to expedite the implementation of essential environmental protections.

This bottom-up strategy is in marked contrast to the top-down bioregional institutions that the federal and state governments promoted in the 1950s and 1960s, intergovernmental ventures that left "political actors free to play their own game without the counterweight of a focused public opinion." Shifting away from this insular orientation has immense appeal to Eric Snyder, planning director of Sussex County, New Jersey, and he decries the "Balkanized land-use decision-making environment" in which he and his colleagues for too long had operated. "We've so many agencies involved and each has its own legislative mandates [and] limits of jurisdiction. It's really difficult for anything other than chaos to come out of it. In Common Waters, we've got some people with the right idea," a more nimble approach that can "break down some of those barriers."[26]

To nurture such regional collaborations, the Pinchot Institute has raised money to underwrite the Common Waters Partnership and the eponymous fund. Realizing in the words of former Grey Towers director Edgar Brannon that "the health of our regional economy is very much tied to the quality of the living environment," the fund commenced investing its dollars—to date over seven hundred thousand dollars have been released—as incentives to promote "water friendly" forest management, underwrite the creation of conservation easements to preserve forest cover, and improve the "finances of forest ownership so families can afford to keep their forests as forests." As in Ecuador and the Pacific Northwest, the ambition of the Common Waters project is to sustain the land and the people who make it their home.[27]

The project's impact on the Carr-Dreher Farm in Sterling, Wayne County, Pennsylvania, is a case in point. The small, seventy-nine-acre family-owned forest, sitting on one of the highest points of the Pocono Mountains, is draped over two ridges and lies within the upper reaches of Butternut Creek. The site's elevation is one factor complicating its restoration: high winds and ice storms periodically damage the upcountry woodlands, already weakened from insect infestations and extensive deer browsing that has eliminated natural regeneration. Moreover, the landscape has been a working bluestone quarry since the early twentieth century. With the death of the family patriarch, ownership passed to his two daughters and their families who were faced with a difficult dilemma—to sell off the land, then bringing in no income, or to restore it without the necessary resources to do so. The county forester, who was linked into the Common Waters project, made it possible for the family to take the path of most beneficence. After a formal assessment of their property's damaged condition, and following a successful application for Common Waters dollars, the Carr-Drehers were able to hire a consulting forester to develop a stewardship plan, seek additional funding and tax relief from a variety of county, state, and federal entities to clear away rock and logging debris and invasive species in hopes of encouraging the regeneration of the native maple, beech, and cherry tree forest. For all the satisfaction that family has gained in repairing the land and its integrity one acre at a time, it has also derived satisfaction knowing that these environmental gains are not theirs alone. Their commitment upstream, in the words of Gary Carr, has given "the gift of guaranteed clean water to the millions downstream in the Delaware Valley whom we will never know or meet."[28]

That same vision has motivated the Jorritsma family, owners of the historic Ideal Farm in Lafayette Township, Sussex County, New Jersey, to collaborate with Common Waters and a host of associated agencies and organizations to regenerate its overgrazed riparian ecosystem. Once the site of a major dairy operation that sprawled across 3,400 acres, Ideal Farm's prize-winning, purebred Guernsey herd of two thousand produced milk marketed throughout the New York metropolitan area. Although the family scaled back the dairy in the 1960s, heavy grazing since the early twentieth century already had taken its toll, badly compromising a long stretch of land that straddles Paulins Kill, a Class 1 tributary of the Delaware. The browsing animals had trampled wetlands and had stripped away almost all trees and woody vegetation, leaving behind "a well-established herbaceous layer" but no "beneficial overstory shade cover" and minimal "vegetative biodiversity resulting in marginal wildlife habitat." The stream is in no better condition—without shade, water temperatures in the summer have spiked up to 70 degrees, making it difficult for aquatic species to survive. The banks, lacking root stabilization, are "vertical or near vertical for several thousand feet," as the stream meanders through the property, and are subject to "sloughing due to undercutting [and] frost heaving." As for the floodplain, wet agricultural acres, and dedicated cropland, these too have been experiencing intensified damage from more frequent flooding—a once-fertile ground was in dire straits.[29]

Its distressed state makes it an ideal candidate for restoration. After extensive water-quality analyses and site surveys, in 2012 the Wallkill River Watershed Management Group, with support from the Pinchot Institute and Common Waters, the New Jersey Department of Environmental Protection, Suffolk County agencies and the Nature Conservancy, and the U.S. National Conservation Resource Service and the Fish and Wildlife Service, started work on the ten-acre site. Professional ecologists and volunteers removed an array of invasive species and planted seedlings, cuttings, and container stock throughout the project area on both sides of the Paulins Kill. Willows, sycamores, and swamp oak, tulip poplar, pin oak, and dogwoods have been staked in the stream banks, sunk in the wetlands, and dug into the floodplain and cropland. It will take time to regenerate a terrain that has been under considerable stress since 1910, but this initial step is critical for reasons ecological and educational. To rebuild the biological integrity of this section of the Paulins Kill watershed is important in its own right; with the goal of treating the causes of degradation as well as returning habitat complexity to a landscape that cows made simple, Common Waters collaborators have an opportunity to revive what seemed dead. Although it may take more than an individual's lifetime to bring the full length of the Paulins back to its more complex dynamic, that is all the more reason to stitch together as many contributing entities as possible to sustain the larger goal into the future. Jan Jorritsma, whose grandfather founded Ideal Farms in 1910, put it this way: "life is too short not to be cooperative."[30]

Proponents of EcoMadera and the Forest Health-Human Health Initiative could easily echo her claim, an argument that also undergirds much of the Pinchot Institute's activism since the early 1990s. Weaving together different coalitions to meet the diverse needs—technical, environmental, social, and financial—of the residents of Christobal Colon and Vernonia, Oregon, or of those inhabiting the high ground and lowlands giving shape to the Delaware River watershed, has strengthened communities, biotic and human. In one sense that's a far cry from the institute's original 1960s mission framed around the need to advance conservation education in the immediate aftermath of Rachel Carson's *Silent Spring*. Yet its work is also consistent with the larger implications of its earlier aspiration to help Americans comprehend their obligation to enhance the health of people and places, striving to make the planet more habitable and just—a new greatest good for the long run, a moral imperative worth pursuing.

CHAPTER TWELVE

Looking Forward

The past, we are told, is prologue. But that does not mean its prescriptions are always translatable by the present or in the future. When Gifford Pinchot helped galvanize the nation to conserve and sustainably manage its forests or be confronted with a "timber famine," the world population was well under two billion. Six decades later, when President Kennedy spoke at the dedication of the Pinchot Institute at Grey Towers, in which he too invoked a vision of environmental limits and natural resource scarcity, less than three billion people lived on the planet. Not surprisingly, then, the answers that many twentieth-century conservationists developed in response to the varied environmental crises of their times depended for their resolution on an educational process in which adults and children could be taught to appreciate nature as an ecosystem and an amenity, a place of study and play. These treasured, if imperiled landscapes, were in jeopardy in good measure because of our often-thoughtless resource consumption. Even as we learned how our patterns of behavior, individual and collective, too rapidly depleted the very materials on which our life depended, we were schooled by forester Pinchot, the Audubon Society's Mabel Osgood Wright, and John Muir of the Sierra Club, as well as Aldo Leopold, Rachel Carson, and a host of others, that regulatory protections should be enacted to save productive terrain and bio-rich wilderness. With careful scientific analyses, with the right mix of political and economic incentives, we could rein ourselves in.

Certainly that is the prescriptive world within which the Pinchot Institute has operated since 1963, with its practical conservationism wedded to a pragmatic approach to the resolution of some of the nation's nagging environmental problems. Its commitments have long presumed that small-scale change can have landscape-scale implications, that it is possible to repair the world one acre at a time.

It is not obvious that individuals or organizations can operate on any other presumption, but this modus operandi is now facing a decisive test. We will shortly experience a global population of ten billion people, with intensifying needs for food, water, shelter, and energy. In the past, advances in technology have pushed back earlier perceptions of limits, but how much longer that strategy can play out is uncertain, especially as the production of these resources and their conversion into energy are fueling a radical change in the Earth's climate.

Even the slow shift to renewable resources—water, forests, agriculture, non–fossil fuel energy—will complicate existing economic, political, and national security systems worldwide. Another challenge awaiting scientists who study and the people and organizations charged with sustainably managing these resources is that the world is also becoming less predictable in very unsettling ways. The strident yet ultimately sterile debate over whether humanity is responsible for climate change cannot alter the fact that the climate is changing and in ways that already is confounding the management of renewable resources.[1]

Most of what we know about how forests and other ecosystems function, and how they respond to management practices and other human interventions, was developed during the past two centuries during a period of relative stability in the world's climates. Against this backdrop of predictable annual temperature regimes, wet and dry seasons, and their effects on everything from water availability to insect populations, we have built entire bodies of science in ecology, agronomy, and silviculture. Upon these foundations of reliable and widely accepted knowledge, we also have constructed our standard practices for forest management and farming.

Those certainties now are in flux. Forests tell part of this tale: some are experiencing unprecedented, near-total mortality from insects, disease, and wildfires; elsewhere, trees are starting to appear on the tundra, by definition a landscape devoid of the woody. As the optimal habitat zones for particular tree species and their associated ecosystems shift toward higher latitudes or elevations, the existing forests increasingly become relict systems, rooted where they are as their progeny follow climate characteristics that are moving steadily farther away. These relict populations become subject to increasing environmental stresses from drought and temperature. Eventually they succumb to an insect or disease infestation that instead of killing a few susceptible individual trees wipes out tens of millions. These swaths of dead and dying trees, so evident on the slopes of the Rocky Mountains, set the conditions for wildfires of unprecedented scope and intensity. What grows back are not species that were there before, but those more characteristic of warmer and drier climes. In some cases, it is not a forest at all that returns, but grassland or some other biome better adapted to the emerging regional climate.[2]

This evolution is reminiscent of how climates have changed in temperate, boreal, and tropical ecosystems for millions of years—*reminiscent but not the same.* Geologists, paleobotanists, and others whose research focuses on time frames in the thousands or millions of years have documented the ebb and flow of several glacial periods that have had profound effects. Through methods such as the analysis of pollen in core samples drawn from ancient lake sediments, scientists have documented how the surrounding landscape was home to succeeding ecosystems—from tundra to coniferous forest to temperate forest and back again—during interglacial periods lasting from twenty to fifty thousand years, a gradual process in which species migrated, mutated, adapted, or died in response to the slow shift in regional climate.[3]

Therein lies the major difference for natural and human systems in the early twenty-first century. The concentrations of greenhouse gases in the Earth's atmosphere, which will rise in coming decades under even the most optimistic projections for reducing carbon dioxide emissions, indicate that many regions of the globe will experience during the next century or so a degree of climate change that would normally take place over thousands of years. There simply is not enough time for the kinds of ecosystem migration, mutation, and adaptation that has taken place in previous interglacial periods.[4]

Scientists in response have coined a new name for this era of unprecedented and rapid climatological disruption: "the Anthropocene." The proposed term not only demarcates a temporal break from the Holocene, the current interglacial period following the Pleistocene glaciations that ended 25,000 years ago, but as its linguistic root announces it identifies humanity's profound impact on climate as this era's defining force.

This emerging reality, however named, has significant implications for conservation, renewable-resource management, and environmental sustainability. It calls into question, for example, whether we can continue to rely upon "the balance of nature" to set things right—or even whether the very concept of a "balance of nature" can long endure. This idea has been the scientific bedrock for long-standing efforts to restore populations of condors, ospreys, salmon, or wolves to their native habitat. It also has provided the justification for setting aside large parks, reserves, and refuges to protect threatened portions of the planet's biodiversity. But a warming Earth upends this supposition and "undermines almost all the rules that environmental stewards have lived by." These preserves, with fixed boundaries on a map, and the idea of preservation itself, on which the creation of such bounded places rely, will not protect the intended species and ecosystems as the organisms themselves already are responding to changing climates. If we persist in applying a strict preservationist approach to maintaining historic biota in evolving landscapes and marinescapes, that work may prove futile, turning "conservationists into something closer to gardeners and zookeepers."[5]

More nimble and perhaps more successful will be a conservationism that is adapted to the Anthropocene. It will take as a given that ecosystems are increasingly fluid, dynamic, even unstable, environments whose gyrations must also be set in the context of a burgeoning human population putting unprecedented pressure on the land for shelter and sustenance. Proponents of an Anthropocenic-framed conservationism, then, will have to figure out how to protect water resources, wildlife habitat, biodiversity, and other essential resources and services located on public and private lands. Critical too will be their ability to ensure the economic viability of local economies and communities whose futures are inextricably linked with the natural systems they inhabit. Making certain that equity is a key element in these resolutions, a notion too long segregated from discussions about the necessity of and desire for ecosystem preservation, is essential. "The Anthropocene does not represent the failure of environmentalism," argues Emma Marris, "it is the stage on which a new, more positive and forward-looking environmentalism can be built. This is the Earth we have created, and we have a duty, as a species, to protect it and manage it with love and intelligence.[6]

When President Kennedy dedicated the Pinchot Institute neither he nor any of the other citizens gathered there had any way of knowing about these looming challenges in the decades ahead. They had some inkling that conservation as it had been conceived since the late nineteenth century to the mid-twentieth was about to change, that the commodity orientation of much of its work was increasingly out of step with an emerging cultural demand for wilderness preservation, recreation opportunities, and the protection of biodiversity. In truth, it would take several more years before that notion would be more fully realized and embraced, heralded with the passage of such seminal legislation as the Wilderness Act (1964), Wild and Scenic Rivers Act (1968), National Environmental Policy Act (1970), the Clean Air (1970), Clean Water Act (1972), Endangered Species Act (1973), and the National Forest Management Act (1976). It would take some time too before anyone appreciated how these initiatives would power an environmental movement whose authority gained further impetus from the civil rights and feminist movements, an engaged reciprocity that made the late 1960s and early 1970s so transformative.

How appropriate then that in 1962, one year before President Kennedy dedicated the Pinchot Institute, two game-changing books appeared: Rachel Carson's *Silent Spring* and Thomas Kuhn's *The Structure of Scientific Revolutions.* Carson's text, a beautifully crafted critique of the insidious damage that pesticides wreaked on healthy ecosystems, quickly became the touchstone of an emboldened environmental movement. It warnings stimulated a paradigm shift in how Americans thought about their place in and responsibility for the environment.[7]

The very notion of paradigmatic alterations was itself the subject of Kuhn's best-selling study in which he probed how scientific investigations can produce dramatic breaks in the way that scientists conceive of the world. Such alterations do not happen immediately, for discoveries that run counter to conventional wisdom are often resisted at first. At some point the growing body of empirical evidence forces scientists to rethink what they know and how they know it. That is the trigger for paradigmatic breaks, a revolutionary and disruptive jolt—much as *Silent Spring* shook up American political culture. "Led by a new paradigm, scientists adopt new instruments and look in new places. Even more important," Kuhn asserted, they come to "see new and different things when looking with familiar instruments in places they have looked before. It is rather as if the professional community had been transported to another planet where familiar objects were seen in a different light and are joined by unfamiliar ones as well."[8]

The late 1960s was one such moment, not least for conservationists for whom the language of the environment and environmentalism upset once widely accepted notions of how to articulate and manage the human place within the natural world. Contemporary debates within the Forest Service and the fledgling Pinchot Institute mirrored this larger tumult, an intense questioning that now finds its parallel in the early twenty-first century as we argue over the impact that climate change will have on the prospects for life on earth.

Up for grabs too is the role that conservationism might play in mitigating and adapting to shifting climes. Contemporary ecologists and conservationists take passionate issue with one another over what the ubiquitous influence of humanity on natural ecosystems means for the future of conservation. Some state flatly that biodiversity conservation as currently practiced is failing; others caution against running up the white flag of surrender, not wanting to suggest that there no longer is any point worrying about conservation. We may be witnessing the kind of intellectual collision that presages the next paradigm shift.[9]

If so, we cannot predict its outcome. But whatever set of ideas emerge as the most salient and persuasive, conservation in the Anthropocene still will depend on multidisciplinary investigations to produce research that will shape policy and practice. It will require an open, participatory, and transparent civic arena in which these ideas can be explained and vetted. And it must revolve around a combination of public and private initiatives and incentives to ensure their enactment, from the ground up. Ultimately the success of this new conservationism will hinge on and be measured by the degree to which its proponents' actions are collaborative, equitable, sustainable, and just.[10]

Or, as Gifford Pinchot argued in 1907: because it is "the duty of the people to think and act for the benefit of the whole people," conservation implies, indeed "demands the application of common sense to common problems for the common good." Some principles may well be timeless.[11]

NOTES

Introduction

1. Stephanie Pendergass, "The Common Waters Fund: A Forest-to-Faucet Approach for the Delaware River," *Pinchot Letter* 16, no. 1 (Spring 2011): 1–3.

2. Nathaniel G. Sajdak to Char Miller, July 26, 2012.

3. Char Miller, *Gifford Pinchot and the Making of Modern Environmentalism* (Washington, D.C.: Island Press, 2001).

4. John F. Kennedy, "Remarks of the President at Pinchot Institute for Conservation Studies, Milford, Pennsylvania, September 24, 1963," http://www.foresthistory.org/ASPNET/Places/GreyTowers/JFK_speech.pdf (accessed May 9, 2013).

5. Interview with William Price, Pinchot Institute for Conservation, December 10, 2012; Gifford Pinchot, "Grazing in the Forest Reserves," *Forester*, November 1901, 276.

6. V. Alaric Sample, "Moderation in Defense of Democracy Is No Vice," *Pinchot Letter* (Fall 2011): 3; the article's title is also a tempering of conservative Senator Barry Goldwater's assertion that "extremism in the defense of liberty is no vice." By the nature of its self-representation, the Pinchot Institute is also appealing to a particular audience and set of funders, philanthropic or contractual. About the boundary-setting quality of such appeals generally, Thomas Medvetz argues suggestively that think tanks have a "common need for political recognition, funding, and media attention. These needs powerfully limit the think tank's capacity to challenge unspoken premises of policy debate, to ask original questions, and to offer policy prescriptions that run counter to the interests of financial donors, politicians, or media institutions" (Thomas Medvetz, *Think Tanks in America* [Chicago: University of Chicago Press, 2012], 7).

7. Kennedy, "Remarks of the President at Pinchot Institute for Conservation Studies."

Chapter 1. This Old House

Epigraph: Gifford Pinchot quoted in the *Port Jervis Union-Gazette,* June 10, 1933, Media Binder, Grey Towers National Historic Site (hereafter GTNHS).

1. "Pilgrimage to Milford Marks Pinchot Anniversary," *Pennsylvania Forests* 41, no. 3 (Fall 1961), Media Binder, GTNHS; H. R. Frantz, "Chapter Suggests Pinchot Memorial," *Journal of Forestry* 59, no. 11 (November 1961), 844.

2. Frantz, "Chapter Suggests Pinchot Memorial," 844.

3. Gifford Pinchot, *Fishing Talk* (Harrisburg, PA: Stackpole Books, 1993), 233–39.

4. Peter Pinchot, "Remarks by Peter Pinchot on the 30th Anniversary of the Dedication of Grey Towers to the U.S. Forest Service," *Conservation Legacy* 9 (September 25, 1993), http://www.foresthistory.org/ASPNET/Places/GreyTowers/30th_Anniversary.pdf (accessed May 9, 2013); undated interview with Sarah H. (Sally) Pinchot, GTNHS; Adam Rome, *The Bulldozer in the Countryside: Suburban Sprawl and the Rise of American Environmentalism* (New York: Cambridge University Press, 2001).

5. Gifford B. Pinchot to Lt. Colonel Barber, November 19, 1960; December 9, 1960, Folder 313.47A, GTNHS; August 23, 1961; October 17, 1961, November 27, 1961, Folder 313.47B, GTNHS; "Col. Barber to Speak," *Schenectady Gazette,* February 15, 1963, http://news.google.com/newspapers?nid=1917&dat= 19630215&id= EggrAAAAIBAJ&sjid=OokFAAAAIBAJ&pg=2394,1942099 (accessed May 9, 2013).

6. Gifford B. Pinchot to E. Carleton Granbury, April 26, 1961, Folder 313.47B, GTNHS; Gifford B. Pinchot to Mrs. Amos R. E. Pinchot, June 28, 1961, Folder 313.47B, GTNHS.

7. Gifford B. Pinchot to E. Carleton Granbury, April 26, 1961; Gifford B. Pinchot to A. S. Fabian, May 9, 1961; June 8, 1961; Gifford B. Pinchot to William Hinkle, May 9, 1961; Gifford B. Pinchot to J. Pennington Straus, June 8, 1961, Folder 313.47B, GTNHS.

8. Gifford B. Pinchot to Richard E. McArdle, September 8, 1961, Folder 313.47B, GTNHS; Harold B. Borneman, "The Pinchot Era," in Gifford B. Pinchot file, GTNHS. In 1950 Averell Harriman and his brother Roland had given Arden, the family estate in Harriman, New York, to Columbia University, and it became the location of the American Assembly, a public policy institute that Dwight D. Eisenhower created; like Grey Towers, the site was named a National Historical Landmark in 1966.

9. Arthur W. Greeley to Gifford B. Pinchot, December 22, 1961, Folder 313.47B, GTNHS.

10. Ibid.

11. Ibid.

12. Ibid; Samuel H. Ordway Jr. to Richard E. McArdle, March 7, 1962, Folder 313.47C, GTNHS.

13. Samuel H. Ordway Jr. to Richard E. McArdle, March 7, 1962; Matthew J. Brennan to William B. LaVenture, March 19, 1962; Richard R. Hadley to Gifford B. Pinchot, April 18, 1962; Richard R. Hadley to J. Pennington Straus, April 18, 1962, Folder 313.47C, GTNHS.

14. Edward Cliff to Samuel Ordway, May 2, 1962, Folder 313.47C, GTNHS.

15. George E. Brewer to Gifford B. Pinchot, May 4, 1962; Gifford B. Pinchot to George E. Brewer, May 8, 1962, Folder 313.47C, GTNHS.

16. This Gifford Pinchot was the child of Amos and his first wife Gertrude Minturn and named for his famous uncle. Laughing about the family's naming

pattern, Gifford B. Pinchot declared: "There are a bunch of us—Dad, my cousin Giff (known as 'Long Giff,' because he is tall and thin and I am short and squat), my son Gifford and his son Gifford. This shows that we don't have much imagination about first names in the family," Gifford B. Pinchot, memorial exhibit material, GTNHS.

17. Matthew Brennan to Gifford B. Pinchot, undated (internal evidence indicates it was written in March 1962); William B. LaVenture to Harold Bornemann, August 31, 1962 (with draft memorandum); Clint Davis to Gifford B. Pinchot, November 6, 1962 (with draft of Clint Davis to Mrs. Amos Pinchot, November 6, 1962); Gifford B. Pinchot to Clint Davis, November 13, 1962; memorandum: December 26, 1962, Folder 313.47C, GTNHS.

18. "Should Be Memorial Park," *Pike County Dispatch,* November 22, 1962, Media Binders, GTNHS; see chapter 3 for a more complete discussion of the Pinchot family's gifts to Milford; Nelson C. Brown to Gifford B. Pinchot, May 3, 1962; Gifford B. Pinchot to Nelson C. Brown, June 1, 1962, Folder 313.47C, GTNHS.

19. Norman B. Lehde, "US Conservation Center Scheduled for Milford," *Port Jervis Union-Gazette,* May 24, 1963; "Milford: Birthplace of American Forestry," *Pike County Dispatch,* May 24, 1963; Norman B. Lehde, "Along Milford's Streets," *Pike County Dispatch,* May 25, 1963; unsigned editorial, *Pike County Dispatch,* May 30, 1963; Jacqueline Depuy, "Plans for Pinchot Estate Seen as Future Heritage," *Port Jervis Union-Gazette,* June 10, 1963; Howard MacDonald, "Pinchot Home, Living Tribute to Former Owner," *Pocono Times Herald Record,* June 21, 1963, Media Binders, GTNHS.

20. Norman B. Lehde, "Pinchot Institute Will Honor Forestry Pioneer," *Port Jervis Union-Gazette,* September 23, 1963; reprinted in Norman B. Lehde, ed., *When President Kennedy Visited Pike County* (Milford, PA: Pike County Chamber of Commerce, 1964), 15.

Chapter 2. September 24, 1963

1. Norman B. Lehde, *When President Kennedy Visited Pike County* (Milford, PA: Pike County Chamber of Commerce, 1964), 18–19.

2. Mike Sparks, "Grey Towers," *Dixie Ranger* 34, no. 3 (November 2003): 5–7; Jack Gooden, "Whatever Happened to the Pinchot Institute?," *Eastern Region Retirees Newsletter,* August 2001, 11–12; Jack Godden to Gerald W. Williams, June 6, 2001, U.S. Forest Service Collection, R/F box 14, Forest History Society; Gerald W. Williams to Jack Godden, June 19, 2001, U.S. Forest Service Collection, R/F box 14, Forest History Society; Jack Godden to S. Forney, June 9, 2004, U.S. Forest Service Collection, R/F box 14, Forest History Society; "Behind the Scenes for the JFK Visit," *Union Gazette,* September 24, 1963; Norman Lehde, "Along Milford's Shaded Streets," *Pike County Dispatch,* September 21, 1963, Media Binders, Grey Towers National Historic Site (hereafter GTNHS); F. Dale Robinson, "Rededication of Pinchot Institute for Conservation," September 25, 1963, F7.3, Forest History Society.

3. Lehde, "Along Milford's Shaded Streets."

4. Bibi Gaston to Char Miller, June 3, 2012; a more complete discussion of Gaston's impressions that day is in Bibi Gaston, *The Loveliest Woman in America: A Tragic Actress, Her Lost Diaries and Her Granddaughter's Search for Home* (New York:

William Morrow, 2008), 168–71; "Remarks by Peter Pinchot on the 30th Anniversary of the Dedication of Grey Towers to the U.S. Forest Service," *Conservation Legacy* 9 (September 25, 1993), http://www.foresthistory.org/ASPNET/Places/GreyTowers/30th_Anniversary.pdf, last (accessed May 9, 2013); interview with Sally Pinchot, undated, GTNHS.

5. Dan Dwyer, "The Day JFK Was Here," *Port Jervis Union-Gazette,* September 25, 1963.

6. Ibid.; Norman B. Lehde, "JFK's Visit Thrills Thousands," *Pike County Dispatch,* undated, Media Binders, GTNHS. Because Governor William Scranton had not played any role in the creation of the Pinchot Institute, and had no ceremonial role as did the president in its acceptance on behalf of the nation, I did not include his brief remarks in the chapters that follow.

Chapter 3. Home Grounds

Epigraph: Cornelia Bryce Pinchot quoted in Amy L. Snyder, "Grey Towers National Historic Landmark: Recreating a Historic Landscape" (master's thesis, Cornell University, 1988), 23.

1. Gifford Bryce Pinchot, "Remarks," Folder 149.25, Grey Towers National Historic Site (hereafter GTNHS).

2. Some of the following paragraphs draw off my previous writing about the Pinchot family's history in Char Miller, *Gifford Pinchot and the Making of Modern Environmentalism* (Washington, D.C.: Island Press, 2001), 15–34; "Edgar Pinchot," in *Commemorative Biographical Record of Northeastern Pennsylvania* (Chicago: T. H. Beers, 1900), 277; and in Alfred Mathews, *A History of Wayne, Pike and Monroe Counties, Pennsylvania* (Philadelphia: R. T. Peek, 1886), 862–63; Daniel Resnick, *The White Terror and the Political Reaction after Waterloo* (Cambridge: Harvard University Press, 1966); Guillaume de Bertier de Sauvigny, *The Bourbon Restoration* (Philadelphia: University of Pennsylvania Press, 1967), 102–7, 117–19, 134–35; Gifford Pinchot, *Breaking New Ground* (New York: Harcourt, Brace, 1947), 10.

3. Mathews, *A History of Wayne, Pike and Monroe Counties,* 860–61.

4. J. Hector St. John de Crevecoeur, *Letters from an American Farmer* (New York: Penguin Books, 1981), 71; Gay Wilson Allen and Roger Asselineau, *St. John de Crevecoeur: The Life of an American Farmer* (New York: Viking Penguin, Inc., 1987), 32–45; Mathews, *A History of Wayne, Pike and Monroe Counties,* 860–64; William Bross, quoted in Mathews, *A History of Wayne, Pike and Monroe Counties,* 883–86.

5. John Brodhead to Cyrille C. D. (C. C. D.) Pinchot, August 9 and 30, 1839; John Wallace to C. C. D. Pinchot, April 15 and May 6, 1834; Wallace to C. C. D. Pinchot, May 5, 1835; Seth Couch to C. C. D. Pinchot, February 9, 1836; John Brodhead to C. C. D. Pinchot, May 19, 1837; R. R. Boughton to C. C. D. Pinchot, September 18, 1837, Gifford Pinchot Papers, Library of Congress; Thomas R. Cox, "Transition in the Woods: Log Drivers, Raftsmen, and the Emergence of Modern Lumbering in Pennsylvania," *Pennsylvania Magazine of Biography and History* 104, no. 3, July 1980, 345–64; Robert K. McGregor, "Changing Technologies and Forest Consumption

in the Upper Delaware Valley, 1780–1880," *Journal of Forest History* 32, no. 2 (April 1988): 69–81.

6. Thomas R. Cox, *The Lumberman's Frontier: Three Centuries of Land Use, Society, and Change in America's Forests* (Corvallis: Oregon State University Press, 2010), 73–100; Gordon S. Whitney, *From Coastal Wilderness to Fruited Plain: A History of Environmental Change in Temperate North America From 1500 to the Present* (Cambridge: Cambridge University Press, 1995); Michael Williams, *Americans and Their Forests: A Historical Geography* (New York: Cambridge University Press, 1989), 146–89; Thomas Cox et al., *This Well-Wooded Land: Americans and Their Forests from the Colonial Times to the Present* (Lincoln: University of Nebraska Press, 1985), 110–90; McGregor, "Changing Technologies," 69–81; James Elliot Defenbaugh, *A History of the Lumber Industry of America*, 2 vols., (Chicago: American Lumberman, 1906–1907), 563–64.

7. Mathews, *A History of Wayne, Pike and Monroe Counties,* 863; "Cyrille Pinchot" file, GTNHS; Carol Severance, "The American Art Collection of James Pinchot," seminar paper, Cooperstown Graduate Programs, Fall 1993, 1–2, copy on file at GTNHS.

8. "Edgar Pinchot," *Commemorative Biographical Record of Northeastern Pennsylvania,* 277; Mathews, *A History of Wayne, Pike and Monroe Counties,* 863–65, 868, 892–93; "Edgar Pinchot," *Commemorative Biography of Northeastern Pennsylvania,* 270, 277, 369, 580; Mott, *From the Ocean to the Lakes,* 344–45; Kristen Doran on-camera interview with Gifford Pinchot III, April 1, 2012, WVIA-PBS, Scranton, PA.

9. John Weir to James Pinchot, May 17, 1870; James Pinchot to C. C. D. Pinchot, September 24, 1871; James Pinchot to Eliza Pinchot, August 21, 1871; James Pinchot to C. C. D. Pinchot, June 16, 1864; James Pinchot to C. C. D. Pinchot, December 30, 1869, Gifford Pinchot Papers, Library of Congress; Richard L. Bushman, *The Refinement of America: Persons, Houses, Cities* (New York: Knopf, 1992), 238–79, 313–401.

10. Matthews, *A History of Wayne, Pike and Monroe Counties,* 863–64; David Chase, "Superb Privacies: The Later Domestic Commissions of Richard Morris Hunt, 1878–1895," *The Architecture of Richard Morris Hunt,* ed. Susan R. Stern (Chicago: University of Chicago Press, 1986), 164–65; Paul R. Baker, *Richard Morris Hunt* (Cambridge: MIT Press, 1980), 337–40; Bushman, *The Refinement of America,* 353–401.

11. Mary Eno Pinchot to James Pinchot, August 21, 1888, Gifford Pinchot Papers, Library of Congress; Kristen Dorn on-camera interview with Peter Pinchot, March 16, 2012, WVIA-PBS, Scranton, PA.

12. "An Address of Gifford Pinchot, Esq., Delivered on Center Square, August 28, 1889, to Commemorate the Second Centennial of the Republic," *Milford Dispatch,* August 29, 1889, GTNHS.

13. Miller, *Gifford Pinchot and the Making of Modern Environmentalism,* 177–81.

14. Ibid., 301–8.

15. Gifford B. Pinchot, *Loki and Loon: A Lifetime Affair with the Sea* (New York: Dodd, Mead and Company, 1985), 2–4, 128; interview with Sarah H. Pinchot, GTNHS. Shortly after his return from the extensive voyage, Gifford Bryce penned *Giff and Stiff in the South Seas* (Philadelphia: John C. Winston Company, 1933) about his

adventures; he later put its royalties to good use after his 1935 marriage with his wife Sally, buying their first sailboat; it launched their collective love of seagoing, which they pursued for the next fifty years.

16. David Piece, "Long Arm of Government Changed the Poconos Forever," *Pocono Record,* August 12, 2001; Kathleen Duca-Sandberg, "The History and Demise of the Tocks Island Dam Project: Environmental War or the War in Vietnam," (PhD diss., Seton Hall University, 2011), http://scholarship.shu.edu/cgi/viewcontent.cgi?article=1009&context=dissertations (accessed May 9, 2013).

17. Miller, *Gifford Pinchot and the Making of Modern Environmentalism,* 379.

18. "The Blow That Probably Wouldn't Have Killed His Father," *American Forests,* editorial, May 1973, 9; "Protest from Dr. Pinchot," *American Forests,* November 1973, 2; Al Weiner, "Gifford Pinchot Would Have Laughed," *American Forests,* November 1973, 12–13, 34–37; a more complete analysis is in Miller, *Gifford Pinchot and the Making of Modern Environmentalism,* 357–61.

19. Cornelia Bryce Pinchot, "Gifford Pinchot and the Conservation Ideal," *Journal of Forestry* 49, no. 2 (February 1950): 86.

Chapter 4. The Inseparable World

Epigraph: Samuel H. Ordway Jr., *Prosperity beyond Tomorrow* (New York: Roland, 1955), viii.

1. Remarks by Sam Ordway at the Pinchot Institute for Conservation Studies, September 24, 1963, Grey Towers National Historic Site (hereafter GTNHS).

2. George Perkins Marsh, *Man and Nature, or, Physical Geography as Modified by Human Action* (Seattle: University of Washington Press, 2000), 328–29.

3. Aldo Leopold, *Game Management* (New York: Charles Scribner's Sons, 1933), 26, 172; Thomas Robertson, "Total War and the Total Environment: Fairfield Osborn, William Vogt, and the Birth of Global Ecology," *Environmental History* 17 (April 2012): 336–64.

4. Elwood Maunder, *Conservation's Communicator: An Oral Interview with Henry Clepper* (Santa Cruz, CA: Forest History Society, 1976), 49–50; Henry Osborn, *Our Plundered Planet* (Boston: Little, Brown, 1948); William Vogt, *The Road to Survival* (New York: William Sloan and Associates, 1948).

5. Robertson, "Total War and the Total Environment," 336–64; see also Daniel Horowitz, *Anxieties of Affluence: Critiques of American Consumer Culture, 1939–1979* (Amherst: University of Massachusetts Press, 2004), 20–47; Lisbeth Cohen, *A Consumer's Republic: The Politics of Mass Consumption in the Postwar America* (New York: Random House, 2008); John Kenneth Galbraith, *The Affluent Society* (New York: Houghton Mifflin, 1958).

6. Gifford Pinchot, "A Forest-Devastation Warning," in *Pan-American Cooperation in Forestry Conservation* (Washington, D.C.: Government Printing Office, 1925), 5–9; Gifford Pinchot, "Conservation as a Foundation for Permanent Peace," *Nature,* August 10, 1940, 183–85; Char Miller, *Gifford Pinchot and the Making of Modern Environmentalism* (Washington, D.C.: Island Press, 2001), 16–17, 361–65.

7. Gifford Pinchot, *Breaking New Ground* (New York: Harcourt, Brace, 1947), 504–10.

8. Samuel H. Ordway Jr., "Plunder or Plenty," *Saturday Review*, April 15, 1961, 13–15.

9. Osborn, *Our Plundered Planet*, 163, 20; Fairfield Osborn, *The Limits of the Earth* (Boston: Little, Brown and Company, 1953); and Fairfield Osborn, ed., *Our Crowded Planet: Essays on the Pressures of Population* (New York: Doubleday, 1962) update his immediate postwar arguments; Samuel H. Ordway Jr., *Resources and the American Dream, including a Theory of the Limit of Growth* (New York: Roland Press, 1953), 5–7; Ordway, *Prosperity beyond Tomorrow*, 178.

10. Osborn, *Prosperity beyond Tomorrow*, 26, 33–34.

11. Ibid., 40.

12. Robert Gottlieb, *Forcing the Spring: The Transformation of the American Environmental Movement*, revised edition (Washington, D.C.: Island Press, 2005), 151–52.

13. Robin W. Winks, *Laurance Rockefeller: Catalyst for Conservation* (Washington, D.C.: Island Press, 1997); on Paul Hale, see Maunder, *Conservation's Communicator*, 50–51; as the environment movement moved to more public demonstrations and radical proclamations, even benign Earth Day activism seemed beyond the pale (Gottlieb, *Forcing the Spring*, 151–52).

14. Gifford B. Pinchot, *Loki and Loon: A Lifetime Affair with the Sea* (New York: Dodd, Mead and Company, 1985), 14–15.

15. Ibid.; Gifford B. Pinchot, Remarks at the Dedication of the Pinchot Institute for Conservation Studies, September 24, 1963, GTNHS.

16. Gifford Bryce Pinchot, "Whale Culture: A Proposal," *Perspectives in Biology and Medicine* 10, no. 1 (August 1966): 33–43; "Marine Farming," *Scientific American* 223, no. 6 (September 1970): 15–21; "Ecological Aquaculture," *Bioscience* 24, no. 5 (1974): 265; Samuel H. Ordway Jr., *A Conservation Handbook* (New York: Conservation Foundation, 1949), 55–56; Ordway, *Prosperity beyond Tomorrow*, 178–79; Orville L. Freeman, *World without Hunger* (New York: Frederick A. Praeger, 1968), x: "We are dealing with a race between what could be done and what will be done, as much as with a race between population and food supply. We already have the knowledge and the resources to achieve the desired balance between food and people. What is lacking is the willingness to an all-out attack on hunger." For the Secretary of Agriculture, U.S. overseas food aid was simply a Band-Aid, and as such, "the only possible long-term answer" was clear—"the bulk of the world's food must be produced where it is to be consumed."

17. Gottlieb, *Forcing the Spring*, 70–76; Robertson, "Total War and the Total Environment," 336–64.

Chapter 5. Under Fire

Epigraph: Remarks by Secretary of Agriculture Orville L. Freeman, introducing President John F. Kennedy at the dedication of the the Pinchot Institute for Conservation Studies, Milford, Pennsylvania, September 24, 1963, Grey Towers National Historic Site (hereafter GTNHS).

1. Orville L. Freeman, recorded interview by Charles T. Morrissey, December 15, 1964, John F. Kennedy Library Oral History Program; Arthur M. Schlesinger

Jr., *A Thousand Days: John F. Kennedy in the White House* (Boston: Houghton Mifflin Company, 1965), 64, 119, 144.

2. Freeman, recorded interview, John F. Kennedy Library Oral History Program.

3. Remarks by Secretary of Agriculture Orville L. Freeman, introducing President John F. Kennedy; Edward P. Cliff to Regional Foresters, Directors, and Area Directors, April 28, 1971, GTNHS: this letter, which conveyed a copy of the agriculture secretary's foundational letter to Pinchot in 1905, was to be inserted "in the history files in all headquarters offices. . . . The original letter, framed, now hangs in [Cliff's] office. More than a prized memento, it is at the very heart of Forest Service history."

4. Remarks by Secretary of Agriculture Orville L. Freeman, introducing President John F. Kennedy.

5. Ibid.

6. Cornelia B. Pinchot, "Gifford Pinchot and the Conservation Ideal," *Journal of Forestry* (February 1950): 83–86; Char Miller, *Gifford Pinchot and the Making of Modern Environmentalism* (Washington, D.C.: Island Press, 2001).

7. Char Miller, "Crisis Management: Challenge and Controversy in Forest Service History," *Rangelands* (June 2005): 14–18. In his Grey Towers speech, Secretary Freeman alluded to the 1905 land transfer, gingerly, because the national forests were then located in the Department of the Interior (which was not then "the great conservation organization it is today under the leadership of Stewart Udall. Instead it was primarily a land 'disposal agency'"). Pinchot concluded that the "thing to do . . . was to get some of the outstanding areas under different jurisdiction where they could be properly conserved. He went to Theodore Roosevelt. Roosevelt enthusiastically agreed. Together they convinced Congress to enact legislation to transfer the existing Federal forests to the Department of Agriculture."

8. Char Miller, *Public Lands, Public Debates: A Century of Controversy* (Corvallis: Oregon State University Press, 2012), 16–35.

9. Miller, *Public Lands, Public Debates,* 79–90; Kristen Doran on-camera interview with Peter Pinchot, March 16, 2012, WVIA-PBS, Scranton, PA; Hamlin Garland, *Cavanaugh: A Forest Ranger; or Romance in the Woods* (New York: Harper and Brothers, 1910).

10. Char Miller, "Crisis Management," 14–18.

11. Martin Nie, "The Bitterroot Controversy," http://forestryencyclopedia.jot.com/ WikiHome/Bitterroot%20Controversy (accessed June 12, 2012).

12. Ronald B. Hartzer and David A. Clary, eds., *Half a Century in Forest Conservation: A Biography and Oral History of Edward P. Cliff* (Washington, D.C.: USDA Forest Service, 1981); Harold K. Steen, *The Chiefs Remember: The Forest Service, 1952–2001* (Durham, NC: Forest History Society, 2004), 26–39.

13. Ibid.

14. Steen, *The Chiefs Remember,* 26–39; *Report of the Chief of the Forest Service, 1964,* quoted in Miles Burnett and Charles Davis, "Getting Out the Cut: Politics and National Forest Timber Harvests, 1960–1995," *Administration and Society* 34, no. 5 (May 2002): 206; Jack Ward Thomas, "What Now? From a Former Forest Service Chief," in *A Vision for the Forest Service: Goals for Its Next Century,* ed. Roger Sedjo

(Washington, D.C.: Resources for the Future, 2000), 10–43; Char Miller, *Public Lands, Public Debates*, 151–55.

Chapter 6. Greening the Presidency

Epigraph: John F. Kennedy, "Remarks to the White House Conference on Conservation, May 25, 1962," *The American Presidency Project*, Gerhard Peters and John T. Woolley, http://www.presidency.ucsb.edu/ws/print.php?pid=8684 (accessed May 9, 2013).

1. Benjamin C. Bradlee, *Conversations with Kennedy* (New York: W. W. Norton, 1975), 212–14; in a later interview about that auspicious day in Milford, Toni Bradlee recalled the president's interaction with her and her sister: "He was easy with both us. There was no sexual thing evident. I always felt he liked me as much as Mary. You could say there was a little rivalry" (quoted in Sally Bedell Smith, *Grace and Power: The Private World of the Kennedy White House* [New York: Random House, 2004], 410–11); Toni Bradlee in another context admitted that as a senator the president had made a pass at her, which she rebuffed; Mary Meyer was murdered under suspicious circumstances in October 1964, see Nina Burleigh, *A Very Private Woman: The Life and Unsolved Murder of Presidential Mistress Mary Meyer* (New York: Bantam Books, 1998), 218–19; Bibi Gaston, *The Loveliest Woman in America: A Tragic Actress, Her Lost Diaries, and Her Granddaughter's Search for Home* (New York: William Morrow, 2008), 167–70.

2. Bradlee, *Conversations with Kennedy*, 212–14; tracking the differing political positions of the various Pinchot family members is Char Miller, *Gifford Pinchot and the Making of Modern Environmentalism* (Washington, D.C.: Island Press, 2001), 223–24, 244–47, 255–56, 372–73.

3. Bradlee, *Conversations with Kennedy*, 214; Gifford Bryce Pinchot interpreted the scene more benignly: "One thing that impressed me very much about President Kennedy was, in spite of the pressures on him, he was very human. . . . My aunt is a supporter of Senator Goldwater. The president and my aunt were posed by photographers on her front porch and after a picture was taken, he turned to her with a broad and very friendly grin and said, 'Mrs. Pinchot, I will send you a copy of this and you can send it on to Barry" (Norman B. Lehde, *When President Kennedy Visited Pike County* [Milford, PA: Pike County Chamber of Commerce, 1964], 38).

4. Bradlee, *Conversations with Kennedy*, 212–13.

5. Ibid.; Smith, *Grace and Power*, 410–11.

6. Arthur M. Schlesinger Jr., *A Thousand Days: John F. Kennedy in the White House* (Boston: Houghton Mifflin, 1965), 658–60.

7. Brian Clark Howard, "The Ten Greenest Presidents in U.S. History," *The Daily Green*, http://www.thedailygreen.com/environmental-news/latest/greenest -presidents-460808 (accessed May 9, 2013); Howard also published concurrently a listing of the "Nine U.S. Presidents with the Worst Environmental Records," ranging from William B. McKinley (as the best of the worst) to George W. Bush (the nadir), http://www.thedailygreen.com/environmental-news/latest/presidents-worst -environmental-records-460808 (accessed May 9, 2013).

8. John F. Kennedy, "Remarks of the President at Pinchot Institute for Conserva-

tion Studies, Milford Pennsylvania, September 24, 1963," http://www.foresthistory
.org/ASPNET/Places/GreyTowers/JFK_speech.pdf (accessed May 9, 2013); Schlesinger, *A Thousand Days,* 658–60.

9. Kennedy, "Remarks of the President at Pinchot Institute for Conservation
Studies."

10. Samuel Hays, *Conservation and the Gospel of Efficiency: The Progressive Conservation
Movement 1890–1920* (New York: Athenaeum, 1959); Kathleen Duca-Sandberg, "The
History and Demise of the Tocks Island Dam Project: Environmental War or the
War in Vietnam" (master's thesis, Seton Hall University, 2011), http://scholarship.shu.
edu/dissertations/30 (accessed May 9, 2013).

11. Richard Reeves, *President Kennedy: Profile of Power* (New York: Simon and
Schuster, 1993), 605–7; John F. Kennedy to Gaylord Nelson, May 16, 1963; Nelson to
Kennedy, May 24, 1963; Nelson to Kennedy, August 29, 1963, Wisconsin Historical
Society, http://www.nelsonearthday.net/collection/conservation-tour.htm (accessed
May 9, 2013); offering the most complete analysis of Kennedy's conservation agenda
is Thomas G. Smith, "John Kennedy, Stewart Udall, and New Frontier Conserva-
tion," *Pacific Historical Review* 64, no. 3 (August 1995): 329–66.

12. Smith, "John Kennedy, Stewart Udall, and New Frontier Conservation," 352–55.

13. John F. Kennedy, "Remarks at the Convention Center in Las Vegas, Nevada,"
September 28, 1963, *The American Presidency Project,* Gerhard Peters and John T.
Woolley, http://www.presidency.ucsb.edu/ws/?pid=9443 (accessed May 9, 2013);
"Remarks at the High School Memorial Stadium, Great Falls, Montana," September
26, 1963, *The American Presidency Project,* Gerhard Peters and John T. Woolley, http://
www.presidency.ucsb.edu/ws/?pid=9435 (accessed May 9, 2013), in which Kennedy
promoted the regional tourism industry: "we have here in the western United States
a section of the world richer by far almost than any other. I want them to come
out here. And I want the United States to take those measures in this decade which
will make the Northwest United States a garden to attract people from all over this
country and all over the world"; "Remarks at the Hanford, Washington, Electric
Generating Plant," September 26, 1963, *The American Presidency Project,* Gerhard
Peters and John T. Woolley, http://www.presidency.ucsb.edu/ws/?pid=9436
(accessed May 9, 2013), in which Kennedy reiterated science's capacity to transform
human life: "But the other part of conservation is the newer part, and that is to use
science and technology to achieve significant breakthroughs as we are doing today,
and in that way to conserve the resources which 10 or 20 or 30 years ago may have
been wholly unknown. So we use nuclear power for peaceful purposes and power.
We use new techniques to develop new kinds of coal and oil from shale, and all the
rest. We use new techniques that Senator Magnuson has pioneered in oceanography,
so from the bottom of the ocean and from the ocean we get all the resources which
are there, and which are going to be mined and harvested. And from the sun we are
going to find more and more uses for that energy whose power we are so conscious
of today. All this means that we put science to work, science to work in improving
our environment and making this country a better place in which to live"; Kennedy's
speech at the Mormon Tabernacle was the only one of the fifteen he delivered on

his tour that did not speak to conservation issues and was entirely driven by matters of faith: "For it is a harsh fact that we have tended in recent times to neglect these deeper values in favor of our material strength. We have traveled in 100 years from the age of the pioneer to the age of payola. We boast to foreign visitors of our great dams and cities and wealth but not our free religious heritage. We have become missionaries abroad of a wide range of doctrines—free enterprise, anticommunism, and pro-Americanism—but rarely the doctrine of religious liberty" ("Speech of Senator John F. Kennedy, Salt Lake City, Utah, Mormon Tabernacle," September 23, 1960, *The American Presidency Project,* Gerhard Peters and John T. Woolley, http://www.presidency.ucsb.edu/ws/?pid=74176 [accessed May 9, 2013]).

14. Duca-Sandberg, "The History and Demise of the Tocks Island Dam Project"; Rachel Carson to Steward Udall is quoted in Smith, "John Kennedy, Stewart Udall, and New Frontier Conservation," 348.

15. Bradlee, *Conversations with Kennedy,* 212–14; John F. Kennedy to Gaylord Nelson, May 16, 1963; Nelson to Kennedy, May 24, 1963; Nelson to Kennedy, August 29, 1963, Wisconsin Historical Society, http://www.nelsonearthday.net/collection/conservation-tour.htm (accessed May 9, 2013); Senator Nelson's disappointment with the conservation tour is captured in Gaylord Nelson, Susan M. Campbell, and Paul A. Wozniak, *Beyond Earth Day: Fulfilling the Promise* (Madison: University of Wisconsin Press, 2002); Bill Christopherson, *The Man from Clear Lake: Earth Day Founder Senator Gaylord Nelson* (Madison: University of Wisconsin Press, 2004), 175–86; Gerald S. and Deborah H. Strober, *"Let Us Begin Anew": An Oral History of the Kennedy Presidency* (New York: HarperCollins Publishers, 1993), 248; other evaluations of the presidential trip's impact are Schlesinger, *A Thousand Days,* 658–60; Reeves, *President Kennedy,* 605–7; Smith, "John Kennedy, Stewart Udall, and New Frontier Conservation," 351–57.

16. Kennedy, "Remarks of the President at Pinchot Institute for Conservation Studies"; several months earlier, in his introduction to Stewart Udall's *The Quiet Crisis,* the president also had voiced his concern about access to and the equitable distribution of resources, aligning himself with long-standing progressive discontent with the antidemocratic thrust of corporate monopolization of nature: "George Perkins Marsh pointed out a century ago that greed and shortsightedness were the natural enemies of a prudent resources policy. Each generation must deal anew with the 'raiders,' with the scramble to use public resources for private profit, and with the tendency to prefer short-run profits to long-run necessities. The nation's battle to preserve the common estate is far from won" (Stewart Udall, *The Quiet Crisis* [New York: Holt, Rinehart and Wilson, 1963], xi); Charles H. W. Foster, *The Cape Cod National Seashore: A Landmark Alliance* (Hanover, NH: University Press of New England, 1985).

17. Kennedy, "Remarks of the President at Pinchot Institute for Conservation Studies."

18. Kennedy, "Remarks to the White House Conference on Conservation"; Smith, "John Kennedy, Stewart Udall, and New Frontier Conservation," 340–43, 361–62.

Chapter 7. Conservation Education

Epigraphs: Maurice K. Goddard, then Pennsylvania secretary of forests and waters, is quoted in "Pine Island Jetport Outlook Held Good," *Port Jervis Union-Gazette,* June 19, 1963, Media Binder, Grey Towers National Historical Site (hereafter GTNHS); Matthew J. Brennan, codirector of the Pinchot Institute, is quoted in "Two Dedicated Men to Direct Institute," unsourced clipping, September 1963, Media Binders, GTNHS.

1. John F. Kennedy, "Remarks of the President at Pinchot Institute for Conservation Studies, Milford, Pennsylvania, September 24, 1963," http://www.foresthistory.org/ASPNET/Places/GreyTowers/JFK_speech.pdf (accessed May 9, 2013).

2. Norman B. Lehde, "JFK's Visit Thrills Thousands," *Pike County Dispatch,* undated, Media Binders, GTNHS.

3. Gifford Bryce Pinchot to John Gray, May 13, 1982, Appendix D-II, Grey Towers Master Site and Interpretative Plan (Pinchot Institute for Conservation Studies, 1982), GTNHS; Pinchot to Max Peterson, April 21, 1986; Pinchot to Barry Walsh, November 14, 1986; Pinchot to Edward J. Vandermillen, November 21, 1986, U.S. Forest Service Collection, F8.2, Forest History Society; Clint Davis to Gifford Bryce Pinchot, November 6, 1962, Pinchot to Davis, November 13, 1962, Folder 313.47C, GTNHS; Gifford Bryce Pinchot to Russell E. Train, March 20, 1967, Folder 313.49B, GTNHS; Matthew Brennan, Monthly Report, June 4, 1965, Folder 149.11, GTNHS.

4. Ruth Pinchot to Matthew Brennan, April 12, 1964, Folder 149.19, GTNHS.

5. Edward Cliff to Orville Freeman, June 22, 1964; Ordway to Cliff, February 6, 1964, Folder 149.5, GTNHS.

6. Rounding out the board was E. DeAlton Partridge, president of Montclair State University; George A. Selke, former dean of the University of Montana forestry school; and Wilson Clark at the Montana College of Education; "Press Release," November 17, 1964, Folder 149.5, GTNHS.

7. Matthew J. Brennan, "Conservation for Youth," paper presented at the annual meeting of the Massachusetts Audubon Society, February 16, 1957; Matthew J. Brennan, "Total Education for the Total Environment," paper presented to the American Association for the Advancement of Science (1964), reprinted in the *Journal of Environmental Education* 6, no. 1 (1974): 16–19; Matthew J. Brennan, "Conservation as an Area of Study Appropriate to Science," *Science Teacher* 34, no. 4 (April 1967): 16–20; *Forest Service Directory, 1960–1968,* Forest History Society; "Two Dedicated Men to Direct Institute," unsourced, undated clipping, September 1963, Media Binders, GTNHS; Matthew Brennan to Pike County Chamber of Commerce, February 28, 1964, Folder 149.19, GTNHS; Matthew J. Brennan, *A Teacher's Guide to the Conservation of People and Their Environment* (Port Washington, NY: Bayberry Press, 1984).

8. Calvin Stillman, "The Pinchot Institute at Grey Towers," Folder 149.17, GTNHS; after his retirement, Stillman served as volunteer archivist at Grey Towers, and

I owe a debt to him for his significant contribution to the site's archival organization and the commentary he wrote about some of its key holdings.

9. Matthew J. Brennan to Pike County Chamber of Commerce, February 28, 1964, Folder 149.19, GTNHS; "Two Dedicated Men to Direct Institute."

10. "Two Dedicated Men to Direct Institute"; Stillman, "The Pinchot Institute at Grey Towers."

11. Louis L. Gould, *Lady Bird Johnson and the Environment* (Lawrence: University Press of Kansas, 1988), 212–14; Lady Bird Johnson, *A White House Diary* (New York: Holt, Reinhart and Winston, 1970), 392; Irene Moore, "Accent on Youth," *American Forests* 72, no. 8 (August 1966): 22–25, 55; Robin W. Winks, *Laurance S. Rockefeller: Catalyst for Conservation* (Washington, D.C.: Island Press, 1997), 142–45. Rockefeller was the chair of the May 1965 White House Conference on Natural Beauty, the prelude to the subsequent youth conference; his relationship with the First Lady, and President Johnson's commitments, according to Winks, meant the chief executive "moved quickly" to implement these paired conferences' shared recommendations.

12. Matthew J. Brennan, "Speech . . . at Pinchot House at Planning Meeting of Youth Representatives . . ." January 29, 1966, F9.4, Forest History Society; Brennan, "Director's Report," Fall 1966, Folder 149.8, GTNHS.

13. Ibid.; Gould, *Lady Bird Johnson and the Environment*, 212–14; Moore, "Accent on Youth," *American Forests*, 22–25, 55; Matthew J. Brennan, "Director's Report," Fall 1966, Folder 149.8, GTNHS; Laurance S. Rockefeller, *A Report to the Nation: National Youth Conference on Natural Beauty and Conservation, Washington, D.C., June 26–29, 1966* (New York: National Youth Conference on Natural Beauty and Conservation, 1967), S2.6.1.

14. Brennan, "Director's Report," Folder 149.8, GTNHS.

15. Paul F. Brandwein, "Creativity and Personality in the Scientist," in *Rethinking Science Education: The Fifty-ninth Yearbook of the National Society for the Study of Education,* ed. Nelson B. Henry (Chicago: National Society for the Study of Education, 1960), 63–81; Paul F. Brandwein, *The Gifted Child as Future Scientist: The High School Student and His Commitment to Science* (New York: Harcourt, Brace and Company, 1955); Paul F. Brandwein, "Origins of Public Policy and Practice in Conservation: Early Education and the Conservation of Sanative Environments," in *Future Environments of North America,* eds. Frank Fraser Darling and John P. Milton, (Garden City: Natural History Press, 1966), 629–47; Paul F. Brandwein, "Conservation," *Science Teacher* 34, no. 4 (April 1967): 13; Gifford Bryce Pinchot to Russell E. Train, April 7, 1967, Folder 313.49B, GTNHS; Charles E. Roth, "Paul F. Brandwein and Conservation Education," in *One Legacy of Paul F. Brandwein,* 91–93, http://books.google.com/books?id=813_FZknf FcC&pg=PA92&lpg= PA92&dq=the+legacy+-+national+friends+of+grey+towers&source=bl&ots= eZzyzFKXoH&sig=FB7XzxbMIYXo4seHP_Hs7PUNKZU&sa=X&ei=66ouUPuQ Nqi9yAGrqYCYBQ&ved=oCBYQ6AEwADge#v= onepage&q=abeles&f=false (accessed May 9, 2013).

16. Matthew J. Brennan, "Speech . . . at Pinchot House at Planning Meeting of

Youth Representatives . . ."; "Techniques of Teaching Conservation," summary of the October 12–14, 1966 conference (USDA-Forest Service and the Conference Foundation, December 14, 1967, GTNHS; Summary of Major Recommendations Made at the Conference on the Future of Conservation Education, August 22–25, 1965, (USDA Forest Service and the Conservation Foundation, August 1966), GTNHS.

17. Matthew J. Brennan, "Monthly Report," September 7 and October 3, 1964, Folder 149.19, GTNHS; February 2, 1965; June 4, 1965; October 1, 1965, Folder 149.11, GTNHS; "Director's Report," February 2, 1967, Folder 149.8, GTNHS.

18. J. Brooks Flippen, *Conservative Conservationist: Russell E. Train and the Emergence of American Environmentalism* (Baton Rouge: Louisiana State University Press, 2006), 43–46.

19. Ibid., 43–46; Char Miller, *Gifford Pinchot and the Making of Modern Environmentalism* (Washington, D.C.: Island Press, 2001), 226–28; Mark Harvey, *Wilderness Forever: Howard Zahniser and the Path to the Wilderness Act* (Seattle: University of Washington Press, 2005), chapter 9.

20. "Pinchot Institute Curtails Program," *Port Jervis Union-Gazette,* February 1, 1967; see also Deborah C. Fort, ed., *One Legacy of Paul F. Brandwein,* Classics in Science Education, 91–93, http://books.google.com/books?id=813_FZknfFcC& pg=PA92&lpg=PA92&dq=the+legacy+-+national+friends+of+grey+towers& source=bl&ots=eZzyzFKXoH&sig=FB7XzxbMIYXo4seHP_Hs7PUNKZU&sa= X&ei=66ouUPuQNqi9yAGrqYCYBQ&ved=0CBYQ6AEwADge#v=onepage&q =abeles&f=false (accessed May 9, 2013).

21. Matthew J. Brennan to Wilson Clark, May 31, 1967, Folder 149.21, GTNHS; Richard H. Pough to Russell Train, April 12, 1967, Folder 149.21, GTNHS; Stillman, "The Pinchot Institute at Grey Towers," 8, GTNHS;

22. Russell E. Train to board of governors, May 17, 1967; Gifford Bryce Pinchot to Russell E. Train, May 17, 1967, Folder 149.21, GTNHS; Train to Pinchot, March 3, 1967, Folder 313.49B, GTNHS, reveals that Train doubted the institute's capacity to serve as a "*national* conservation program and center" because it was too tied to Paul Brandwein's conceptions of the nature of conservation education—and Train did not believe that Brandwein, or the institute, had garnered the necessary national acceptance for their approach. Given Train's ambition to create a national platform in Washington, Brandwein had to go. So, apparently, did the Pinchot Institute.

23. Board of Governors' Minutes, June 29, 1967, Folder 149.8, GTNHS.

24. Ibid.

25. Ibid.; Samuel P. Hays, "From Conservation to Environment: Environmental Politics in the United States Since World War II," in *Out of the Woods: Essays in Environmental History,* eds. Char Miller and Hal K. Rothman (Pittsburgh: University of Pittsburgh Press, 1997), 101–5; Michael McCloskey, *In the Thick of It: My Life in the Sierra Club* (Washington, D.C.: Island Press, 2005), 236–39.

26. Ibid.

27. Ibid.; Edward P. Cliff and Russell E. Train to David M. Heyman, June 25, 1958, Folder 149.8, GTNHS; identical letters were sent to all members of the board of governors, the advisory committee; a revised release to the media.

28. Russell E. Train, *Politics, Pollution, and Pandas: An Environmental Memoir,* (Washington, D.C.: Island Press, 2003); Flippen, *Conservative Conservationist;* Harold K. Steen, interview with Russell E. Train, 1993, Forest History Society; *Conservation Education in the Forest Service,* November 1999, http://www.fs.usda.gov/Internet/ FSE_DOCUMENTS/fsmrs_100523.pdf (accessed May 9, 2013).

Chapter 8. Branching Out

Epigraph: Jack Ward Thomas and Richard M. DeGraaf, "Raccoons on the Roof," in *Gardening with Wildlife: A Complete Guide to Attracting and Enjoying the Fascinating Creatures in Your Backyard* (Washington, D.C.: National Wildlife Federation, 1974), 68.

1. Jack Ward Thomas and Ronald A. Dixon, "Cemetery Ecology," *Natural History* 82, no. 3 (March 1973): 61.

2. Ibid., 62.

3. Ibid., 66–67.

4. Ibid.

5. Statement of Orville L. Freeman, Secretary of Agriculture, on S. 2036, Before the Subcommittee on Employment and Manpower of the Senate Committee on Labor and Public Welfare on June 20, 1960; Youth Conservation Act of 1959, U.S. Senate, 86th Congress, 1st Session, Calendar No. 533, Report 536; Gerald Dicerbo, "The Legislative History of the Youth Conservation Corps," box 1, U.S. Forest Service Collection, Forest History Society.

6. Warren T. Doolittle, "Research in Urban Forestry," *Journal of Forestry* 67 (September 1969): 650–52, 656.

7. Ibid.

8. Ibid.

9. Edward P. Cliff, "Trees and Forests in the Human Environment," in *Symposium Proceedings: Trees and Forests in an Urbanizing Environment,* eds. Silas Little and John H. Noyes (Amherst: Monograph Series no. 17, University of Massachusetts Cooperative Extension, 1971), 17–21; Glenn Sandiford and Lee P. Herrington, *Pinchot Consortium for Environmental Studies: A Lesson in Cooperation* (Broomhall, PA: Northeastern Forest Experiment Station, 1989), 2–3; Jack Ward Thomas to Char Miller, July 20, 2012; July 13, 2011; Char Miller to Jack Ward Thomas, July 13, 2011 in author's possession.

10. Jack Ward Thomas, David P. Worley, Joseph C. Mawson, Elwood L. Shafer Jr., and Robert W. Wilson Jr., *The Pinchot Institute System for Environmental Forestry Studies,* General Technical Report NE-2 (Upper Darby, PA: Forest Service, Northeastern Forestry Experimental Station, 1972), 51–60.

11. Ibid.

12. Ibid., 1–3.

13. Brian R. Payne and Jack Ward Thomas, "New Developments in Environmental Research," *Proceedings of the Tree Wardens, Arborists, and Utilities Conference* (Salem, MA: Forest Service, 1981), 27–28, 31; Robert Marshall Ricard, "Politics and Policy Processes in Federal Urban Forest Policy Formation and Change" (PhD diss., University of Massachusetts, 2009), 235–42.

14. Gifford Pinchot, for one, had decried arborists and the like as amateurs, dis-

tinguishing them from the professional experts he believed should dominate forestry issues in the United States; Char Miller, "Amateur Hour: Nathaniel H. Egleston and the Forestry Movement in Post-Civil War America," *Forest History Today* (Spring-Fall 2005): 20–26; Char Miller, *Ground Work: Conservation in American Culture* (Durham, NC: Forest History Society, 2007), 21–25; Robert M. Ricard, "Shade Trees and Tree Wardens: Revising Urban Forest History," *Journal of Forestry* 103, no. 5 (2005): 230–33; Ricard, "Politics and Policy Processes," 82, 104–5, 119–20; Payne and Thomas, "New Developments in Environmental Research," 27–28, 31.

15. Eldon W. Ross, *History of the Northeast Research Station: 1973–1998,* General Technical Report 249 (Upper Darby, PA: U.S. Department of Agriculture, Forest Service, Northeastern Research Station, 1998), 11; Susan R. Schrepfer, Edwin van Horn Larson, and Elwood R. Mauder, *A History of the Northeastern Forest Experiment Station: 1923 to 1973,* General Technical Report, NE-7 (Upper Darby, PA: U.S. Department of Agriculture, Forest Service, Northeastern Forest Experiment Station, 1973), 38–39.

16. The Weeks Act (1911), which allowed the Forest Service to purchase private property from willing sellers to protect and regenerate forested watersheds, is an indication of how the agency first conceived of its role upstream; see Char Miller, "Neither Crooked Nor Shady: The Weeks Act and the Virtue of Eastern National Forests, 1899–1911," *Journal of the Theodore Roosevelt Association* 30, no. 4 (Fall 2012): 15–24; Thomas R. Wellock, *Preserving the Nation: The Conservation and Environmental Movements, 1870–2000* (Wheeling, IL: Harlan Davidson, Inc., 2007), 110–11, 171, 180–81.

17. William E. Sopper and Louis T. Kardos, eds., *Recycling Treated Municipal Wastewater and Sludge Through Forest and Cropland* (University Park: Pennsylvania State University Press, 1973).

18. Harold K. Steen, "An Interview with Jack Ward Thomas," May 10–12, 2001, 5–6, http://www.foresthistory.org/Research/Thomas%20JW%20oHI%20Final.pdf (accessed May 9, 2013).

19. Sandiford and Herrington, *Pinchot Consortium for Environmental Studies: A Lesson in Cooperation,* 3; Henry D. Gerhold, "History and Goals of METRIA, the Metropolitan Tree Improvement Alliance," *Journal of Arboriculture* 4, no. 3 (March 1978): 62–66.

20. *Children, Nature, and the Urban Environment: Proceedings of a Symposium-Fair,* General Technical Report NE-30 (Upper Darby, PA: Northeast Research Station -Forest Service-U.S. Department of Agriculture, 1977), iii–v, http://nrs.fs.fed.us/pubs/gtr/gtr_ne30/gtr_ne30.pdf (accessed May 9, 2013).

21. Yi-Fu Tuan, "Experience and Appreciation," in *Children, Nature, and the Urban Environment,* 1, 4–5; Paul Shepard, "Place and Human Development," in *Children, Nature, and the Urban Environment,* 12.

22. Edwin M. Fitch and John F. Shanklin, *The Bureau of Outdoor Education* (New York: Praeger Publishers, 1970), 79–83, 179–81; Samuel T. Dana, *Education and Outdoor Recreation* (Washington, D.C.: Government Printing Office, 1968).

23. George H. Moeller, "Research Priorities in Environmental Education," in *Children, Nature, and the Urban Environment,* 218–22. The Forest Service returned

to the theme of children and nature nearly four decades later with its launch of a new program dubbed Kids in the Woods. It was developed in response to yet another disconcerting analysis of the impact of urbanization and technology on America's youth. In *Last Child in the Woods: Saving Our Children from Nature-Deficit Disorder* (New York: Algonquin Books, 2005), Richard Louv has argued that as "the young spend less and less of their lives in natural surroundings, their senses narrow, physiologically and psychologically, and this reduces the richness of human experience." More troubling are the corresponding spikes in childhood obesity, depression, and attention-deficit disorders, ailments that Louv believes are intimately linked to the number of hours kids spent in front of computers and televisions, in the enclosed and insulated built landscape that keeps nature at bay. The only corrective was through full-scale immersion in and reengagement with mother earth, and he pounded the drum for a consciously planned set of programs in schools and within families that would teach the young "a better way to live with nature." In Louv's claims the Forest Service, which since the 1950s had become increasingly fearful that an urbanized population was less and less likely to understand and support its work, found additional confirmation of its long-standing anxieties. Hence the 2007 Kids in the Woods program, which offered a $500,000 challenge grant distributed to community-based initiatives that had the best chance of combating, in the words of Forest Service Chief Gail Kimbell, the "troubling declines we see in the mental and physical health of our children." The initial round of funding went to twenty-four organizations matching the agency's assessment criteria. Successful applicants, Kimbell observed, "focused on underserved and urban youth; recreation and conservation education," an orientation that would enable them "to inspire future conservation leaders, who can perpetuate the critical role nature and forests play in the quality of life for Americans," language that was an unconscious evocation of the hopes that had been voiced at the 1975 Pinchot Consortium conference on children and nature. See Louv, *Last Child in the Woods;* Richard Louv, "No Child Left Inside," *Orion Magazine,* March/April 2007, http://www.orionmagazine.org/index. php/articles/article/240/, last accessed May 9, 2013; Gail Kimball, "Kids in the Woods: Making the Connection," speech to the Recreation Exchange, Washington, D.C., August 27, 2007, http://www.fs.fed.us/news/2007/speeches/08/kids.shtml (accessed May 9, 2013, on a holiday), one way to judge the success of the Kids in the Woods program is that in 2011, Agriculture Secretary Tom Vilsack announced that $1 million would be spent on two Forest Service cost-sharing programs, Kids in the Woods and the Children's Forest (http://www.fs.fed.us/news/2011/releases/04/ mkiw-recipients.shtml, [accessed May 9, 2013]).

24. Robert DiSilvestro, "NWF: 70 Years on the Front Lines of Conservation," *National Wildlife,* February 1, 2006, http://www.nwf.org/News-and-Magazines/ National-Wildlife/News-and-Views/Archives/2006/NWF-70-Years-on-the-Front -Lines-of-Conservation.aspx, (accessed July 29, 2012; Steen, "Interview with Jack Ward Thomas," 9, http://www.foresthistory.org/Research/Thomas%20JW%200 HI%20Final.pdf (accessed July 29, 2012); Joe Strohm, "Create a Mini-Wilderness," *National Wildlife,* April-May, 1973, 4.

25. Mark Wexler, "The Birth of NWF's Habitat Program," *National Wildlife,* January 6, 2010, http://www.nwf.org/News-and-Magazines/National-Wildlife/Gardening/Archives/2010/The-Birth-of-NWFs-Habitat-Program.aspx (accessed May 9, 2013); Jack Ward Thomas, "Habitat Requirements for Suburban Songbirds—A Pilot Study" (PhD dissertation, University of Massachusetts at Amherst, 1973); Jack Ward Thomas and Richard M. DeGraaf, "Non-Game Wildlife Research in Megalopolis," General Technical Report NE-4 (Upper Darby, PA: Northeastern Forest Experiment Station, 1972).

26. Jack Ward Thomas, Robert O. Brush, and Richard M. DeGraaf, "Invite Wildlife to Your Backyard," *National Wildlife,* April-May 1973, 5–16.

27. James D. Davis, "Wildlife in Your Backyard," "Wildlife in an Urbanizing Environment," University of Massachusetts Cooperative Extension Service, June 1974, 175–77; Richard M. DeGraaf and Jack Ward Thomas, "A Strategy for Wildlife Research in Urban Areas," in ibid., 53–56; Jack Ward Thomas, Richard M. DeGraaf, and Joseph C. Mawson, "A Technique for Evaluating Bird Habitat," in ibid., 159–61; Richard DeGraaf and Jack Ward Thomas, "A Banquet for the Birds," *Natural History* 83, no. 1 (1974): p. 40–45.

28. Sandiford and Herrington, *Pinchot Consortium for Environmental Studies: A Lesson in Cooperation,* 30–31.

29. Ibid., 36–38.

30. Ibid., 31, 39–40.

31. Harold K. Steen, ed., *Jack Ward Thomas: The Journals of a Forest Service Chief* (Durham, NC: Forest History Society, 2004), 15–74; Richard M. DeGraff, biography, http://www.fs.fed.us/ne/amherst/staff/degraaf.html (accessed November 13, 2012).

32. Davis, "Wildlife in Your Backyard," 177; concluding on a similar note is Jack Ward Thomas and Richard M. DeGraaf, "Raccoons on the Roof," in *Gardening with Wildlife* (Washington, D. C.: National Wildlife Federation, 1974), 153–68.

Chapter 9. Turning a White Elephant Gold

Epigraph: James Pinchot, "The Yale Summer School of Forestry," *World's Work,* October 1904.

1. Gifford Bryce Pinchot to John Gray, May 13, 1982, in *Grey Towers Master Site and Interpretative Plan* (Pinchot Institute for Conservation Studies, 1982), Appendix D-II, Grey Towers National Historical Site (hereafter GTNHS).

2. Ibid.

3. Ibid.

4. Ibid.

5. George Vilas to Don Bylsma, August 6, 1976, Folder 149.22, GTNHS; these alterations in Grey Towers' status in the Forest Service hierarchy also can be traced through the *Forest Service Directory, 1963–1972,* U.S. Forest Service Collection, Forest History Society.

6. Pinchot Environmental Institute, September 11, 1969, U.S. Forest Service Collection, Forest History Society.

7. Ed Cliff, "Trees and Forests in the Human Environment," in *A Symposium on Trees*

and *Forests in the Human Environment* (Amherst, MA: Cooperative Extension Service, 1971), 20; Warren T. Doolittle, "A New Role for the Pinchot Institute," *Pennsylvania Forests,* Fall 1970, 81.

8. William F. Hasse to Warren T. Doolittle, March 8, 1972; Doolittle to Hasse, March 13, 1972, Folder 149.2, Pinchot Institute for Conservation Studies Papers, GTNHS.

9. I have edited out some of the duplication in the dialogue: "Briefing to Chief and Staff on Grey Towers," February 18, 1976, S6.6.22, U.S. Forest Service Collection, Forest History Society.

10. Robert M. Lake to Rexford A. Resler, March 18, 1976; "A Proposed Plan for the North American Institute for the Future of Forest Environments," S6.6.22, U.S. Forest Service Collection, Forest History Society.

11. For all his success in steering money to his district—and perhaps because of it—in the late 1980s, McDade would be brought up on federal charges for bribery and racketeering, and after an eight-year investigation, which the representative likened to "Chinese water torture," a jury acquitted him on all counts; see http://www.cnn.com/ALLPOLITICS/1996/news/9608/01/mcdade/index.shtml and http://www.cnn.com/ALLPOLITICS/1996/news/9606/10/mcdade/index.shtml (accessed May 9, 2013).

12. Robert M. Lake to Rexford A. Resler, March 18, 1976; "A Proposed Plan for the North American Institute for the Future of Forest Environments"; Resler to Tom Nelson et al., April 14, 1976, in ibid.; *Grey Towers Master Site and Interpretative Plan,* 21–24.

13. Gifford Bryce Pinchot to Ruth Pinchot, October 5, 1977, Folder 313.49C, GTNHS; the awkward and lengthy title of the new project—the North American Institute for the Future of Forest Environments—ultimately was dropped in favor of reviving the original name, the Pinchot Institute for Conservation Studies; *Report on the Proposed Expanded Program for Pinchot Institute for Conservation Studies, 1976–1977,* 22, GTNHS.

14. John L. Gray biography, http://sfrc.ufl.edu/History/gray.html (accessed May 9, 2013).

15. George Vitas to Don Bylsma, "Re: Interpretative Planning for Pinchot Grey Towers VIS Program," August 6, 1976, Folder 149.22, GTNHS; Charles J. Newlon, "Pinchot Institute—Chief and Staff Decisions: The Record," July 26, 1977, Folder 149.22, GTNHS; "Pinchot Institute for Conservation Studies: Activities Completed, Underway and Coming Up in FY 1978," April 5, 1978, Folder 149.15, GTNHS.

16. Preservation of Historic Sites Act, 49 Stat. 666, http://www.cr.nps.gov/history/online_books/anps/anps_3d.htm (accessed May 9, 2013); National Historic Preservation Act, PL 102 575, http://www.nps.gov/history/local-law/nhpa1966.htm (accessed May 9, 2013); William J. Murtagh, *Keeping Time: The History and Theory of Preservation in America,* revised edition (New York: John Wiley and Sons, 1997), 45–47, 68–70; Norman Tyler, *Historical Preservation: An Introduction to Its History, Principles, and Practice* (New York: W. W. Norton, 2000), 44–50; David Hamer, *History in Urban Places: The Historic Districts of the United States* (Columbus: Ohio University Press, 1998), 18–19.

17. David Clary, *Historic Structure Report: Grey Towers,* FS 327 (Washington, D.C.: U.S. Department of Agriculture—Forest Service, 1979), GTNHS; the report was based on the findings of John Milner and Associates, "Final Grey Towers Historic Structures Report, Historic Landscape Report, and Management Plan," GTNHS.

18. Clary, *Historic Structure Report: Grey Towers,* 29–30, 43, 68–75, 77–83.

19. John McGuire to Joanna McGaughey, September 29, 1977, Folder 149.22, GT-NHS; Lora Sharpe, "Milford's Grey Towers Provides Key to Luxury," *Pocono Record,* August 11, 1973; "Grey Towers—Pike County's Camelot," *Pike County Dispatch,* May 13, 1976; "History Relived on Grey Tours Tour," *Wilkes-Barre Times Leader,* July 7, 1977; Monica Von Dobeneck, "Rural Center Breeds City Thoughts," *Pocono Record,* December 26, 1978; Cynthia Van Lierde, "YACC . . . Good Neighbors," *Pike County Dispatch,* May 11, 1978; all articles available in Media Binders, GTNHS.

20. *Report of the Proposed Expanded Program for Pinchot Institute for Conservation Studies,* September 1976 to January 1977, GTNHS; this argument about Grey Tower's vulnerability since 1963 is reaffirmed in the *Grey Towers Master Site and Interpretative Plan,* 1–2, GTNHS.

21. Richard E. McArdle to John Gray, February 28, 1981, in *Grey Towers Master Site and Interpretative Plan,* Appendix D-II, GTNHS; Gifford Pinchot, who Vanderbilt hired in 1891 to cruise upward of 100,000 wooded acres around Biltmore for possible purchase and to establish forest management on his sprawling estate, was troubled by the massive estate and what it represented: "as a feudal castle, it would have been beyond criticism, and perhaps beyond praise. But in the United States of the nineteenth century and among the one-room cabins of the Appalachian mountaineers, it did not belong. The contrast was a devastating commentary on the injustice of concentrated wealth. Even in the early nineties I had sense enough to see that" (Gifford Pinchot, *Breaking New Ground* [New York: Harcourt Brace, 1947], 48.)

22. Richard E. McArdle to John Gray, February 28, 1981, in *Grey Towers Master Site and Interpretative Plan,* Appendix D-II, GTNHS.

23. Ibid.

24. Paul Hirt, *Conspiracy of Optimism: Management of the National Forests Since World War Two* (Lincoln: University of Nebraska Press, 1994), 268–73; James G. Lewis, *The Greatest Good and the Forest Service* (Durham, NC: Forest History Society, 2005), 200–202; Norman B. Lehde, "US Conservation Center Scheduled for Milford," *Port Jervis Union-Gazette,* May 24, 1963, Media Binder, GTNHS.

25. Edgar Brannon to Char Miller, email communication, August 8, 2012; Edgar Brannon, interview, June 26, 2012, Milford.

26. Ibid.; Gary Hines, "Narrative," August 8, 2012; Gary Hines to Char Miller, personal communication, August 8–11, 2012.

27. *Report on the Proposed Expanded Program for Pinchot Institute for Conservation Studies, 1976–1977,* GTNHS, 22; Al Sample to Char Miller, August 12, 2012, personal communication; George Bohlinger to Char Miller, August 12, 2012, personal communication.

28. Gifford Bryce Pinchot to Max Peterson, April 21, 1986, U.S. Forest Service

Collection, F 8.2, Forest History Society; Pinchot to E. J. Vandermillen, November 14, 1986, includes a withering critique of the misinformation that Grey Towers staff was giving out to visitors and a copy of his blistering letter to Barry Walsh, an editor at the Society of American Forests, detailing the many inaccuracies of her article "Grey Towers Celebrates a Centennial," *Journal of Forestry* 84, no. 8 (August 1986): 24–29; Pinchot to Barry Walsh, November 14, 1986; a toned down version of his critique appeared as "A Pinchot Responds," *Journal of Forestry* 85, no. 3 (March 1987): 14. Despite his anger, Pinchot did not believe that Vandermillen was responsible for these error-riddled presentations, scrawling in a postscript: "we're all still very grateful for all you have done at Milford," Gifford Bryce Pinchot Miscellaneous Folder, GTNHS.

29. "Sally Collins: An Interview," *Women in Natural Resources* 23, no. 3 (November 3, 2002), http://www.cnr.uidaho.edu/winr/Collins.htm (accessed May 9, 2013); Daviana D. Apple, "Changing Legal and Social Forests Affecting the Management of National Forests," *Women in Natural Resources* 18, no. 1 (Autumn 1996): 4–10, 40, http://www.cnr.uidaho.edu/winr/Soc_Legal_Forces_ WiNR.pdf (accessed May 9, 2013); Lewis, *The Forest Service and the Greatest Good,* 171–78; Anthony Godfrey, *The Ever-Changing View: A History of the National Forests inCalifornia* (Washington, D.C.: U.S. Department of Agriculture—Forest Service, Pacific Southwest Region, R5-FR-004), 497–505; Edgar Brannon, interview, June 26, 2012.

30. Edgar Brannon, interview, June 26, 2012; Edgar Brannon to Char Miller, August 4, 2012; Carol Severance to Char Miller, August 11 and 13, 2012; Char Miller to Carol Severance, August 11 and 13, 2012; "A First for White Mountains: A Woman Steps into Management," *Nashua Telegraph,* December 9, 1983, 35; David Eugene Conrad and Jay H. Cravens, *The Land We Cared For: A History of the Forest Service's Eastern Region* (Milwaukee, WI: U.S. Department of Agriculture—Forest Service, Region 9, 1997), 272–73; "Grey Towers Director Named," *New Bulletin: State and Private Forestry,* January 15, 1987, Media Binders, GTNHS; Jennifer C. Thomas and Paul Mohai, "Racial, Gender, and Professional Diversification of the Forest Service from 1983 to 1992," *Policy Studies Journal* 23, no. 2 (1995): 296–309.

31. L. L. Burrus-Bammel, "Women and Sexism in Forestry: An Update," *Women in Natural Resources* 11, no. 3 (1989): 23–27; G. Brown and C. Harris, "The Implications of Workforce Diversification in the U.S. Forest Service," *Administration and Society* 25, no. 1 (1993): 85–113; Thomas and Mohai, "Racial, Gender, and Professional Diversification of the Forest Service," 297–98.

32. Carol Severance, "Cornelia Bryce Pinchot (1881–1960)," *History Line* (Spring 1998): 30–36. Amy L. Snyder was another who contributed significantly to this reinterpretation of the Grey Towers master narrative, see her "Grey Towers National Historic Landmark: Recreating a Historic Landscape" (master's thesis, Cornell University, 1988); see also Nancy Pittman Pinchot, a granddaughter of Amos Pinchot and author of "Amos Pinchot: Rebel Prince," *Pennsylvania History* 66, no. 2 (1999): 166–98. She also conducted groundbreaking primary research into the life of actress

Rosamond Pinchot, which later she handed over to her cousin Bibi Gaston for her memoir *The Loveliest Woman in America: A Tragic Actress, Her Lost Diaries, and Her Granddaughter's Search for Home* (New York: William Morrow, 2008).

33. Carol Severance, "The American Art Collection of James W. Pinchot, 1831–1908," (master's thesis, Cooperstown, NY, Graduate Program, 1993); Severance's research deeply informed my analyses of the gender dynamics within the Pinchot family, Char Miller, *Gifford Pinchot and the Making of Modern Environmentalism* (Washington, D.C.: Island Press, 2001), 108–10; Kristin Doran interview with Peter Pinchot, April 1, 2012, WVIA-PBS, Scranton, PA.

34. LeRoy Johnson, "A Review of the Pinchot Institute for Conservation" (January 1992), 8–9, GTNHS.

35. Brannon was intimately involved in the crafting of the Flathead's 1986 forest plan and embroiled in its adjudication, which consisted of thirty-nine separate administrative appeals filed in the following two years; his reflections on the intensity of that work is "Forest Conflict: Learning to Take a Punch," *Pennsylvania Forests* 92, no. 2 (Summer 2001): 15–16; Robert D. Baker et al., *The National Forests of the Northern Region: A Living Legacy* (College Station, TX: Intaglio, 1993), chapter 16, http://www.foresthistory.org/ASPNET/Publications/region/1/history/index.htm (accessed May 9, 2013); Christopher McGory Klyza, *Who Controls the Public Lands? Mining, Forestry, and Grazing Policies, 1870–1990* (Chapel Hill: University of North Carolina Press, 1996), 93–107; Frederick H. Swanson, *The Bitterroot and Mr. Brandborg: Clear-cutting and the Struggle for Sustainable Forestry in the Northern Rockies* (Salt Lake City: University of Utah Press, 2011); Iver Peterson, "50-Year Plan for Forests Renews Debate on Use," *New York Times,* April 12, 1986, 22; Timothy Egan, "Forest Service Abusing Role, Dissidents Say," *New York Times,* March 4, 1990, 1; see Joseph L. Sax and Robert B. Keiter, "Glacier National Park and Its Neighbors: A Study in Federal Relations," *Ecology Law Quarterly* 14, no. 2 (1987): 208–62; John G. Mitchell, "War in the Woods: Swan Song," *Audubon* (November 1989): 92–131; "Environmentalists Win Oil Lease Ruling," United Press International, January 14, 1988; "Mont. National Forest Oil/Gas Leases Jeopardized by U.S. Court Ruling," *Platt's Oilgram News,* March 27, 1985, 3; "Furor on the Flathead," *Newsweek,* September 10, 1984, 25; "Industry Opposes Proposed Montana Land Lockup," *Oil and Gas News,* August 20, 1984, 78; Hirt, *A Conspiracy of Optimism,* 271–78.

36. Edgar Brannon to Char Miller, August 8, 2012; interview with Edgar Brannon, July 13, 2011.

37. Adrian Wooldridge, *Masters of Management: How the Business Gurus and Their Ideas Have Changed the World—for Better and for Worse* (New York: HarperCollins, 2011), 4; John Kotter, *Leading Change* (Cambridge, MA: Harvard Business Review Perss, 1996); Edgar Schein, *Organizational Culture and Leadership* (San Francisco: Jossey-Bass, 1992); Thomas J. Peters and Robert H. Waterman, *In Search of Excellence: Lessons from America's Best Run Corporations* (New York: Grand Central Publishing, 1988); Peter Drucker, *Innovation and Entrepreneurship* (London: Heinemann, 1985); Gifford Pinchot III, *Intrapreneuring: Why You Don't Have to Leave the Corporation to Become an Entrepreneur* (New York: HarperCollins, 1986); Nelson Lichtenstein,

ed., *American Capitalism: Social Thought and Political Economy in the Twentieth Century* (Philadelphia: University of Pennsylvania, 2001); Johnson, "A Review of the Pinchot Institute for Conservation," 9–10; Gifford Pinchot, interview, August 28, 2012; Char Miller, "A Cautionary Tale: Reflections on the Reinvention of the Forest Service," *Journal of Forestry* 94, no. 1 (January 1996): 6–11.

38. Johnson, "A Review of the Pinchot Institute for Conservation," 11–13.

39. *Historic Structures Report: Grey Towers,* 22; Johnson, "A Review of the Pinchot Institute for Conservation," 14–16.

40. E. Florens Rivinus and Mrs. Frederick W. Morris, "A History of the Morris Arboretum," http://www.business-services.upenn.edu/arboretum/archives/HistoricPDF/2011–13–1ArbHistory1975.pdf (accessed May 9, 2013); William Klein's impact on the Morris Arboretum is detailed in the site's application for inclusion on the National Registry of Historic Places (https://www.dot7.state.pa.us/ce_imagery/phmc_scans/H001351_04H.pdf [accessed May 9, 2013]) and in this newspaper assessments of his significance, http://articles.philly.com/1997–02–14/news/25535730_1_morris-arboretum-public-garden-area-gardens; http://articles.philly.com/1990–11–26/news/25928754_1_morris-arboretum-bill-klein-arboretum-director (accessed May 9, 2013); Edgar Brannon, "A Brief History of Grey Towers Restoration, 1990–2005," in author's possession; interview with Edgar Brannon, July 13, 2011, and November 14, 2012; Edgar Brannon to Char Miller, November 14, 2012, email communication; a brief overview of Andropogon's work at Grey Towers is found at http://andropogon.com/grey-towers-national-historic-landmark/ (accessed May 9, 2013); coleaders for the planning effort were Dr. George Skarmeas, head of the Historic Preservation Studio at Vitetta Group, and Carol Franklin, principal and a founding partner of Andropogon.

41. Interview with Edgar Brannon, July 13, 2011, and November 14, 2012; Edgar Brannon to Char Miller, November 14, 2012, email communication; interview with Jack Ward Thomas, July 13, 2011; Edgar Brannon to Char Miller, July 13, 2011.

42. Jack Ward Thomas, "This Time, Our Moment in History, Our Future," unpublished speech, delivered to the Forest Service Leadership Meeting, June 20–23, 1994; Harold K. Steen, ed., *Jack Ward Thomas: The Journals of a Forest Service Chief* (Durham, NC: Forest History Society, 2004), 204, 255, 299; Harold K. Steen, "An Interview with Jack Ward Thomas," May 10–12, 2001, 65, 101, http://www.foresthistory.org/Research/Thomas%20JW%20HI%20Final.pdf (accessed May 9, 2013); interview with Jack Ward Thomas, July 13, 2011; Edgar Brannon to Char Miller, July 13, 2011; interview with Edgar Brannon, November 14, 2012.

43. Edgar Brannon to Char Miller, July 13, 2011; interview with Edgar Brannon, November 14, 2012; Edgar Brannon to Char Miller, November 14, 2012, email communication.

44. Andropogon and Vitetta Group Master Plan, GTNHS; interview with Elizabeth Hawke, June 2004, http://www.foresthistory.org/Events/Folklife/Hawke,%20Elizabeth.pdf (accessed May 9, 2013).

45. "Dr. Gifford B. Pinchot (1915–1989)," *Pike County Dispatch,* Media Binders, GTNHS; Edgar B. Brannon to Susan B. Hess, June 4, 1990; Edward J. Vandermillen

to Mrs. Gifford B. Pinchot, May 30, 1990, U.S. Forest Service Collection, F 7.3, Forest History Society; "Remarks by Peter Pinchot on the 30th Anniversary of the Dedication of Grey Towers to the U.S. Forest Service," *Conservation Legacy* 9 (September 25, 1993), http://www.foresthistory.org/ASPNET/Places/GreyTowers/30th_Anniversary.pdf (accessed May 9, 2013); *Grey Towers National Historic Landmark Strategic Plan, 2000–2005,* 3–5, 9, GTNHS.

Chapter 10. Neutral Site

Epigraph: Hannah J. Cortner, *A More Perfect Union: Democratic and Ecological Sustainability,* Pinchot Institute for Conservation Distinguished Lecture Series, February 19, 2000 (Milford, PA: Grey Towers Press, 2000), 2.

1. *Grey Towers National Historic Landmark, Strategic Plan 2000–2005,* 8, Grey Towers National Historical Site (hereafter GTNHS).

2. Ibid.

3. Rep. Joseph M. McDade, "Other Major Accomplishments," Box 358/7 Grey Towers (1993), Joseph M. McDade Papers, University of Scranton; Joseph McDade, speech at National Friends of Grey Towers, December 8, 1983; Minutes, National Friends of Grey Towers, December 8, 1983; Agenda, National Friends of Grey Towers, February 28, 1984, Box 269/8, Friends of Grey Towers, Joseph M. McDade Papers, University of Scranton.

4. Lee Kimche McGrath to Joseph M. McDade, March 13, 1984, Box 269/8, Joseph M. McDade Papers, University of Scranton.

5. Thomas N. Schenarts to Harry W. Buchanan, February 1985, GTNHS.

6. Paula M. Utermahlen to Debbie Weatherly, May 8, 1984, Box 269/8, Joseph M. McDade Papers, University of Scranton.

7. Ibid.; Sidney L. Krawitz to Debbie Weatherly, April 24, 1984, Box 269/8, Joseph M. McDade Papers, University of Scranton; "McDade Gets Money for Grey Towers in '88 Budget," Box 236/40, Joseph M. McDade Papers, University of Scranton; Edward J. Vandermillen to Joseph M. McDade, April 20, 1987, Folder 149.16, GTNHS.

8. Amy L. Snyder, "Grey Towers National Historic Landmark: Recreating a Historical Landscape" (master's thesis, Cornell University, 1988), 1–16, 21–30, 84–85; Edward J. Vandermillen to Joseph M. McDade, April 20, 1987, Folder 149.16, GTNHS; Edward B. Brannon to Char Miller, September 4, 2012; Gary Hines to Char Miller, September 4, 2012.

9. Ann Christine Reid, ed., *Population Change, Natural Resources, and Regionalism* (Milford PA: Grey Towers Press, 1986), vii–viii; 1–7.

10. John C. Barber and William K. Reilly to Joseph McDade, May 7, 1986, Box 3737/12, Joseph M. McDade Papers, University of Scranton.

11. Reid, *Population Change, Natural Resources, and Regionalism,* vii–viii, 1–7, 47–50, 84–95; Daniel B. Luten, "The Limits-to-Growth Controversy," in *Sourcebook on the Environment: A Guide to the Literature,* eds. Kenneth A. Hammond, George Macinko, and Wilma B. Fairchild (Chicago: University of Chicago Press, 1978), 163–82; on the contemporary debate over regionalism, see also Charles H. W. Foster, *Experiments in*

Bioregionalism: The New England River Basins Story (Hanover and London: University Press of New England, 1984), 173–76; Daniel Press, "Environmental Regionalism and the Struggle for California," *Society and Natural Resources* 8, no. 5 (1995): 288–306; James J. Parsons, "On 'Bioregionalism' and 'Watershed Consciousness,'" *Professional Geographer* 37, no. 1 (February 1985): 1–8.

12. R. Neil Sampson, "Thoughts for the Future," in *Population Change, Natural Resources, and Regionalism*, ed. Ann Christine Reid (Milford PA: Grey Towers Press, 1986), 99–101.

13. Ibid.

14. William E. Shands and Robert G. Healy, *The Lands Nobody Wanted: The Legacy of the Eastern National Forests* (Washington, D.C.: Conservation Foundation, 1977); William E. Shands, V. Alaric Sample, and Dennis LeMaster, *National Forest Planning: Searching for a Common Vision*, volume 2 (U.S. Department of Agriculture—Forest Service, 1990), http://www.fs.usda.gov/Internet/FSE_DOCUMENTS/stelprdb5172341 .pdf (accessed May 9, 2013).

15. Edgar B. Brannon, "Historical Relationship of the Forest Service to the Pinchot Institute for Conservation: A Personal Perspective," January 26, 1996, GTNHS.

16. Chief Robertson met with board members George Bohlinger, an attorney for the Rockefeller brothers, and Ross Whaley, president of the State University of New York College of Environmental Science and Forestry, at that critical 1989 discussion. James Giltmier, "Oral History Interview" (Forest History Society, 2005), 7, 87–90, 97–101, indicates that the organization received $125,000 per year for four years, starting in 1989. Robertson's predecessor at the Forest Service, Max Peterson, then serving on the National Friends board, knew Giltmier from his tenure as a Capitol Hill staffer and had encouraged Giltmier to apply after Giltmier had expressed interest in the position; he had read the job announcement in the *Journal of Forestry* and recalled: "I saw this opening and I thought well, there's something I know something about" (87). See also George Bohlinger to Char Miller, August 11, 2012; Al Sample to Char Miller, August 12, 2012; Edgar Brannon to Char Miller, February 15, 2013; George Bohlinger to Char Miller, February 16, 2013.

17. James W. Giltmier, "Oral History Interview"; Interview with Edgar Brannon, August 22, 2012, and November 14, 2012; Al Sample to Char Miller, August 12, 2012, and September 10, 2012.

18. Frederick H. Swanson, *The Bitterroot and Mr. Brandborg: Clear-cutting and the Struggle for Sustainable Forestry in the Northern Rockies* (Salt Lake City: University of Utah Press, 2011), 268–91.

19. Shands et al., *National Forest Planning*, 5; F. Dale Robertson, "The History of New Perspectives and Ecosystem Management," 3, http://www.srs.fs.usda.gov/ pubs/gtr/gtr_srs074/gtr_srs074-robertson001.pdf (accessed May 9, 2012); Char Miller, *Public Lands/Public Debates: A Century of Controversy* (Corvallis: Oregon State University Press, 2012), 116–21; Samuel P. Hays, *Wars in the Woods: The Rise of Ecological Forestry in America* (Pittsburgh, PA: University of Pittsburg Press, 2007), 17–18, 46, 114–15.

20. Robertson, "The History of New Perspectives and Ecosystem Management,"

3–4; Hays, *Wars in the Woods,* 95; James G. Lewis, *The Forest Service and the Greatest Good: A Centennial History* (Durham, NC: Forest History Society, 2005), 211–16.

21. William E. Shands, Anne Black, and James W. Giltmier, *From New Perspectives to Ecosystem Management: The Report of an Assessment,* Pinchot Institute Monograph Series (Milford PA: Grey Towers Press, 1993), 1–2.

22. Lewis, *The Greatest Good and the Forest Service,* 212.

23. Robertson, "The History of New Perspectives and Ecosystem Management," 6–7.

24. James W. Giltmier, "Foreword," in *From New Perspectives to Ecosystem Management,* eds. William E. Shands, Anne Black, and James W. Giltmier (Milford PA: Grey Towers Press, 1993), n.p., but see also, 22; Hannah J. Cortner and Margaret A. Moote, *The Politics of Ecosystem Management* (Washington, D.C.: Island Press, 1998), 37–52; striking a more hopeful note is Christopher Klyza, *Who Controls the Public Lands? Mining, Forestry, and Grazing Policies, 1870–1990* (Chapel Hill: University of North Carolina Press, 1996), 67–107; he argues that the "technocratic privilege" was on the wane.

25. V. Alaric Sample, Will Price, Jacob S. Donnay, Catherine M. Mater, *National Forest Certification Study: An Evaluation of the Applicability of Forest Stewardship Council (FSC) and Sustainable Forest Initiative (SFI) Standards on Five National Forests* (Pinchot Institute for Conservation, October 22, 2007), 3–4; "Public Review Sought on U.S. National Reports on Sustainable Forestry," *Pinchot Letter* 9, no. 2 (Fall 2004): 1–3; "Independent, Science-Based Report Finds Washington State Forests Could Be Certified as Sustainably Managed," *Pinchot Letter* 8, no. 3 (Fall 2003): 3; "Certification on State Trust Lands (Part 2): Forest Interests Converge in Support of Washington DNR Certification," *Pinchot Letter* 6, no. 2 (Summer 2001): 3–4; *Oregon Forestlands and the Program for the Endorsement of Forest Certification (PEFC): An Assessment of the Process and Basis for Eligibility: Final Report* (Pinchot Institute for Conservation, April 11, 2006); Al Sample, "An Asset Management Approach to Forest Stewardship," part 1, *Pinchot Letter* 7, no. 3 (Winter 2002–2003): 3–5; part 2, *Pinchot Letter* 8, no. 1 (Spring 2003): 5–6; "Certification Pilot Study to Assess Forest Management on Tribal Lands Nationwide," *Pinchot Letter* 6, no. 3 (Winter 2001): 1–2; V. Alaric Sample and Char Miller, "Sustainable Forestry on Tribal Lands, and the Legacy of Gifford Pinchot," *Pinchot Letter* 6, no. 3 (Winter 2001): 8–10.

26. James Snow, "Commentary on the Grey Towers National Historic Site Act," author's possession; James Snow to Char Miller, August 29, 2012; "H.R. 4494—108th Congress: Grey Towers National Historic Site Act of 2004," GovTrack.us (database of federal legislation), http://www.govtrack.us/congress/bills/108/hr4494 (accessed May 9, 2013).

27. Ibid.

28. *Grey Towers National Historic Landmark, Strategic Plan 2000–2005,* 8, GTNHS.

Chapter 11. Common Cause

Epigraph: Catherine Mater interview with Kristin Doran, April 23, 2012, WVIA-PBS, Scranton, PA.

1. Char Miller, *Gifford Pinchot and the Making of Modern Environmentalism* (Washington, D.C.: Island Press, 2001), 55–56; Aldo Leopold, *San County Almanac* (New York: Oxford University Press, 1970), 69; Curt Meine, *Aldo Leopold: His Life and Work* (Madison: University of Wisconsin Press, 1988), 72; John F. Kennedy, "Remarks of the President at the Pinchot Institute for Conservation Studies, Milford, Pennsylvania, September 24, 1963," http://www.foresthistory.org/ASPNET/Places/GreyTowers/JFK_speech.pdf (accessed May 9, 2013).

2. V. Alaric Sample, *Land Stewardship in the Next Era of Conservation* (Milford, PA: Grey Towers Press, 1991), vii–x; Char Miller, *Public Lands, Public Debates: A Century of Controversy* (Corvallis: Oregon State University Press, 2012), 1–5; Harold K. Steen, *The Origins of the National Forests* (Durham, NC: Forest History Society, 1992); John Reiger, *American Sportsmen and the Origins of Conservation* (Corvallis: Oregon State University Press, 2000).

3. Sample, *Land Stewardship in the Next Era of Conservation,* ix, 31–32.

4. Ibid., 32.

5. Peter Pinchot, "Adapting the Legacy of Gifford Pinchot: From Utilitarian to a Communitarian Paradigm," Speech to USDA Forest Service, National Leadership Team, Madison, WI, October 26, 1999.

6. Ibid.

7. "Forest Service Chief Shifts Focus to Clean Water," *Spokane Spokesman-Review,* March 3, 1998, http://m.spokesman.com/stories/1998/mar/03/forest-chief-shifts-focus-to-clean-water-dombecks/ (accessed May 9, 2013).

8. Peter Pinchot, "The Perennial Challenge for Forestry: Sustaining Biological Capital," *Journal of Forestry* 98, no. 11 (November 2000): 64.

9. Peter Pinchot, "Bringing Forestry Back to the First Yale Camp and Bringing the Conservation Debate Back to Grey Towers," *Yale Forestry and Environmental Studies Centennial News* (Fall 1990): 14–16; "The Milford Experimental Forest: A Conservation with Peter Pinchot," *Milford Magazine,* August 2002, 18–19, 21.

10. Pinchot, "Bringing Forestry Back to the First Yale Camp," 14–15.

11. Ibid., 15.

12. Steve Grant, "Blight Fight," *Hartford Courant,* July 11, 2007, http://articles.courant.com/2007–07–11/features/0707110743_1_american-chestnut-foundation-chestnut-trees-connecticut-agricultural-experiment-station; Michael J. Coren, "An American Classic in Yale's Forest," *Yale Alumni Magazine,* November/December 2010, http://www.yalealumnimagazine.com/issues/2010_11/lv_chestnut016.html; "History and Research Converge in American Chestnut Restoration," *USDA Blog,* July 5, 2012, http://blogs.usda.gov/2012/07/27/history-and-research-converge-in-american-chestnut-reintroduction-2/#more-41395 (accessed May 9, 2013).

13. Gifford Pinchot, *A Primer of Forestry* (Washington: Government Printing Office, 1899), 65–66.

14. "Tropical Sustainable Forestry: A New Community Forestry Program in Northwestern Ecuador," *Pinchot Letter* 8, no. 1 (Fall 2003): 1–4.

15. Peter Pinchot, "Demonstrating Sustainable Community Forestry in Ecuador," *Pinchot Letter* 15, no. 3 (Fall 2010): 1, 3–10; in an interview with Kristin Doran (March

16, 2012, WVIA-PBS, Scranton, PA), Peter Pinchot drew an even tighter connection to his grandfather's conception of community-based forestry: "EcoMadera is 100 percent in the mold of the Gifford Pinchot tradition of figuring out how to make forestry work in different environments for the people . . . to make it work economically, and to make it work in the long run for the long-term sustainable management of the forest."

16. Pinchot, "Demonstrating Sustainable Community Forestry in Ecuador," 3–10; Ariel Pinchot, "Connecting Human Health and Forest Conservation in the Rio Verde Canandé Watershed," *Pinchot Letter* 15, no. 3 (Fall 2010): 10–11; "EcoMadera: Market Strategies for Sustainable Forestry," http://www.pinchot.org/gp/EcoMadera (accessed May 9, 2013). Another measure of EcoMadera's viability is whether its strategies are replicable elsewhere, one indicator of which occurred in 2005 when USAID approached Jatun Sacha and the Pinchot Institute to begin working with groups of communities along Ecuador's western border with Columbia and another set of villages in the Amazonian forest. Testifying to its adaptability as well are related projects that have been spun off from the original work at Cristobal Colon. Part of EcoMadera's initial charge has been to study whether it would be possible to restore some of Rio Canandé's once-forested watershed; by 2000, nearly 20 percent of it had been cut over and another 20 percent has been "heavily exploited or is in early stage secondary forest." The Pinchot Institute hired the forest ecologist Amy Rogers to research the factors essential to tropical rainforest regeneration and "to facilitate the development of ecologically based reforestation practices that directly counteract obstacles to forest succession." Among her team's findings is that the inability of primary forest seedlings to (re)establish themselves appears to be a consequence of a "lack of dispersal," a failure that test plots demonstrated could be resolved efficiently and with low labor input through hand dispersal. If this result can be generalized, Pinchot predicted, it "will significantly shorten the time required for the human-mediated restoration of diverse rainforest systems—until now, considered an untenable goal by most tropical foresters" (Amy E. Rogers, "How to Reconstruct a Rain Forest," *Pinchot Letter* 15, no. 3 [Fall 2010]: 14–15; Blair S. Ryearson, "Improving Conventional Timber Harvest and Log Yield in Working Forests of the Mountainous Tropics," *Pinchot Letter* 15, no. 3 [Fall 2010]: 12–13; Pinchot, "Demonstrating Sustainable Community Forestry in Ecuador," 7–9).

17. Ariel Pinchot, "Connecting Human Health and Forest Conservation in the Rio Verde Canandé Watershed," 10–11.

18. Miller, *Gifford Pinchot and the Making of Modern Environmentalism,* 316–19.

19. Interview, April 25, 2012, State Senator Betsy Johnson, WVIA-PBS, Scranton, PA; Tom Hyde quoted in Brian A. Kittler, "Rebuilding a Greener Community through Creative Partnerships," *Pinchot Letter* (Fall 2011): 8–10; Brett J. Butler, "Family Forest Owners of the United States, 2006," General Technical Report, NRS-27 (Newtown Square, PA: U.S. Department of Agriculture—Forest Service, Northern Research Station, 2008), http://www.treesearch.fs.fed.us/pubs/15758 (accessed May 9, 2013).

20. Brett J. Butler, "Family Forest Owners Rule!," *Forest History Today,* Spring/

Fall 2011, 87–91; Catherine Mater, V. Alaric Sample, and Brett J. Butler, "The Next Generation of Private Forest Landowners: Brace for Change," *Pinchot Letter* 10, no. 2 (Winter 2005): 1–4; "Q&A with Catherine Mater," *Pinchot Letter* 10, no. 2 (Winter 2005): 14–16.

21. Catherine Mater, "Linking Forest Health and Human Health in America's Private Woodlands," *Pinchot Letter* (Spring 2009): 11–12.

22. Ibid.; Kittler, "Rebuilding a Greener Community through Creative Partnerships," 8–10; "Cemetery Is Testing Ground for Carbon Valuation," *Daily Astorian,* September 1, 2011, http://m.dailyastorian.com/mobile/article_1f687b50-d4c3-11e0-ae9c-001cc4c03286.html (accessed May 9, 2013).

23. "Cemetery Is Testing Ground for Carbon Valuation"; "The Forest Health-Human Health Initiative," and related survey results, http://www.pinchot.org/gp/FHHHI (accessed May 9, 2013); Catherine Mater to Char Miller, August 2, 2012, email communication.

24. Matthew Brennan, "Speech . . . at Pinchot House at Planning Meeting of Youth Representatives," January 29, 1966, F 9.4, Forest Service Collection, Forest History Society.

25. Rebecca Sanborn Stone and Mary L. Tyrell, "Motivations for Family Forestland Parcelization in the Catskill/Delaware Watersheds of New York," *Journal of Forestry* 110, no. 5 (July/August 2012): 267–74; "UN Water Day: Securing the Future of the Delaware River," *Hunterdon (NJ) County Democrat,* March 23, 2012, http://www.nj.com/hunterdon-county-democrat/index.ssf/2012/03/un_world_water_day_securing_th.html (accessed May 9, 2013).

26. On the relationship between headwater forest management and the Chesapeake Bay, see the special issue of *Pennsylvania Forests* 100, no. 2 (Summer 2009): 18–32, http://www.wbsrc.com/documents/pa%2oforests%2osummer%2oo9.PDF (accessed May 9, 2013); "Common Waters Forum: From the Upper Delaware to the Water Gap," http://www.pinchot.org/gp/Common_Waters_Forum (accessed May 9, 2013); Charles H. W. Foster, *Experiments in Bioregionalism: The New England River Basins Story* (Hanover and London: University Press of New England, 1984), 173–76; Daniel Press, "Environmental Regionalism and the Struggle for California," *Society and Natural Resources* 8, no. 5 (1995): 288–306; James J. Parsons, "On 'Bioregionalism' and 'Watershed Consciousness,'" *Professional Geographer* 37, no. 1 (February 1985): 1–8.

27. That sustainability is being severely tested as energy companies drill countless hydraulic fracturing wells across the Delaware River watershed. Aware that this posed a huge challenge to its restorative work on these same lands, the Pinchot Institute has been active in the regional dialogue that often erupts into debate. "The Marcellus Shale: Resources for Stakeholders in the Upper Delaware Watershed Region," (2011), http://www.pinchot.org/PDFs/Pinchot_Marcellus_Shale_BMPs.pdf (accessed May 9, 2013); "Common Waters Forum: From the Upper Delaware to the Water Gap."

28. Gary Carr to Char Miller, November 8, 2012, email communication; Gary Carr to the Pinchot Institute, June 25, 2012, in author's possession.

29. *Upper Paulins Kill Watershed Restoration Plan, Volume 1,* September 2012, http://

www.wallkillriver.org/uploads/file/cover%20page%20and%20executive%20
summary.pdf (accessed May 9, 2013); "Common Waters Fund Management Practices
Application Project Plan," Wallkill River Watershed Management Group, http://
www.wallkillriver.org/news/article.aspx?id=14 (accessed May 9, 2013).

30. *Upper Paulins Kill Watershed Restoration Plan, Volume 1;* on the principles of
watershed restoration, see Michael P. Dombeck, Christopher A. Wood, and Jack E.
Williams, *From Conquest to Conservation: Our Public Lands Legacy* (Washington, D.C.:
Island Press, 2003), 117–33; Jan Jorritsma to Pinchot Institute Board of Directors,
June 29, 2012.

Chapter 12. Looking Forward

This epilogue is coauthored with V. Alaric Sample, president of the Pinchot Institute,
and draws in part on his essay "Redefining Forest Conservation in the Anthropocene," *Pinchot Letter* 16, no. 4 (Fall 2012): 1–5, http://www.pinchot.org/PICLETTER
FALL12164.pdf (accessed May 9, 2013).

1. John Fedkiw, Douglas MacCleery, and V. Alaric Sample, *Pathway to Sustainability:
Defining the Bounds of Forest Management* (Durham, NC: Forest History Society, 2004).

2. Peter D. Ward, *Under a Green Sky: Global Warming, the Mass Extinctions of the Past,
and What They Can Tell Us About Our Future* (New York: HarperCollins, 2007).

3. Curt Stager, *Deep Future: The Next 100,000 Years of Life on Earth* (New York: St.
Martin's Press, 2011).

4. Ibid.

5. Bill McKibben, *The End of Nature* (New York: Random House, 1989); Bill
McKibben, *Earth: Making a Life on a Tough New Planet* (New York: Times Books, 2011);
Emma Marris, *Rambunctious Garden: Saving Nature in a Post-Wild World* (New York:
Bloomsbury, 2001); Emma Marris et al., "Hope in the Age of Man," *New York Times,*
December 7, 2011, http://www.nytimes.com/2011/12/08/opinion/the-age-of-man
-is-not-a-disaster.html?_r=0 (accessed May 9, 2013); M. Martin Smith and Fiona
Gow, "Unnatural Preservation," *High Country News,* February 4, 2008, http://www
.hcn.org/issues/363/17481 (accessed May 9, 2013); Michelle Mavier, Peter Kareiva,
and Robert Lalasz, "Conservation in the Anthropocene: Beyond Solitude and
Fragility," *Breakthrough Journal,* Winter 2012, http://thebreakthrough.org/index.php/
journal/past-issues/issue-2/conservation-in-the-anthropocene/ (accessed May 9,
2013); Thomas Homer-Dixon, "Our Panarchic Future," *World Watch,* March/April
2009, 8–15.

6. Marris et al., "Hope in the Age of Man," *New York Times,* December 7, 2011,
http://www.nytimes.com/2011/12/08/opinion/the-age-of-man-is-not-a-disaster.
html?_r=0 (accessed May 9, 2013); Mavier, Kareiva, and Lalasz, "Conservation in the
Anthropocene"; Smith and Gow, "Unnatural Preservation."

7. Rachel Carson, *Silent Spring* (Boston, MA: Houghton Mifflin and Company,
1962); on Carson's influence, see Thomas R. Wellock, *Preserving the Nation: The
Conservation and Environmental Movements, 1870–2000* (Wheeling IL: Harlan Davidson
Inc., 2007), 162–65.

8. Thomas S. Kuhn, *The Structure of Scientific Revolutions* (Chicago: University of Chicago Press, 1962), 111.

9. Kareiva, Lalasz, and Marvier, "Conservation in the Anthropocene"; Tim Caro, Jack Darwin, Tavis Forrester, Cynthia Ledoux-Bloom, and Caitlin Wells, "Conservation in the Anthropocene," *Conservation Biology* 26, no. 1 (2011): 185–88; Thomas Lovejoy and Lee Hannah, eds., *Climate Change and Biodiversity* (New Haven: Yale University Press. 2005); Lee Hannah, ed., *Saving a Million Species: Extinction Risk from Climate Change,* (Washington, DC: Island Press, 2011); Marris et al., "Hope in the Age of Man," *New York Times,* http://www.nytimes.com/2011/12/08/opinion/the-age-of-man-is-not-a-disaster.html?_r=0 (accessed May 9, 2014); Mark G. Anderson and Charles E. Ferree, "Conserving the Stage: Climate Change and the Geophysical Underpinnings of Species Diversity," *PloS ONE* 5, no. 7 (2001); http://www.plosone.org/article/info%3Adoi%2F10.1371%2Fjournal.pone.0011554; Mark G. Anderson and A. Olivero Sheldon, *Resilient Sites for Terrestrial Conservation in the Northeast and Mid-Atlantic Region* (The Nature Conservancy, Eastern Conservation Science, 2011), http://conserveonline.org/workspaces/ecs/documents/resilient-sites-for-terrestrial-conservation-1, (accessed May 9, 2010).

10. Richard L. Knight and Courtney White, eds., *Conservation for a New Generation: Redefining Natural Resources Management* (Washington, D.C.: Island Press, 2009), 285–96; Michael P. Dombeck, Christopher A. Wood, and Jack E. Williams, *From Conquest to Conservation: Our Public Lands Legacy* (Washington, D.C.: Island Press, 2003), 163–86.

11. Gifford Pinchot, "Speech to the First National Conservation Congress," August 27, 1907, reprinted in *Conservation,* January 1909, 711–12.

INDEX

Note: Page numbers in italic type indicate illustrations.

Brower, David, 41, 67
Buchanan, Harry, 139, 140
Buck Island Reef, 55
Buckley, William F., 53
Bureau of Outdoor Recreation, 60, 89, 90
Bush, George H. W., 148
Bush, Robert, 91
Butler, Brett, 164

Cape Cod, 59
carbon sequestration, 165, 166
Carr, Gary, *109*, 169
Carr-Dreher Farm, 168–69
Carson, Rachel, 4, 37, 46, 55, 58, 59, 170, 171, 175
Carter, Jimmy, 5, 93
Cavanagh, Forest Ranger (Garland), 48
cemeteries, wildlife habitat in, 78–80
Center for Resource and Environmental Policy, 144
Chicago Wilderness project, 167
Children's Forest, 193n23
civil rights, 54, 80
Clark, Wilson, 73–74, *104*, 188n6
Clary, David, 118, 119
Clean Air Act (1970), 50, 175
Clean Water Act (1972), 50, 175
clear-cutting, 49, 87, 124, 146
Cleveland, Grover, 47
Cliff, Edward P., 18, 23, 65, 66, 67, *102*, *104*; conservation and, 74; environmental education and, 75, 76; Grey Towers and, 46, 113; leadership of, 50, 51
climate change, 172, 173
Clinton, Bill, 93
Cold War, 38, 54
collaboration, 47, 51, 161, 176
Collins, Sally, 125
Common Waters, 2, 166, 168–69, 170; goal of, 167
Common Waters Fund, 2, 109, 167
Common Waters Partnership, 2, 168
Connaughton, Kent, 145
Connecticut Agricultural Research Station, 160

conservation, 5, 25, 41, 43, 54, 58, 59, 68, 119, 174, 186n13; focus on, 16, 57; future of, 176; human, 163; natural resources and, 52; reconceiving, 157; science and, 56, 66; studies, 113; water, 56
"Conservation as a Foundation for Permanent Peace" (Pinchot), 39
conservation education, 36, 43, 63, 75, 76, 89, 112, 137, 139; initiatives for, 68, 69, 90; programming, 66, 67; conservation ethic, 35, 36, 66, 70
Conservation Foundation, 7, 8, 24, 40, 41, 63, 64, 67, 74, 75, 124, 142, 144, 146, 156; commitment by, 72; educational objectives of, 43; Forest Service and, 4, 35, 76, 80, 138; formation of, 37; Grey Towers and, 51, 65; Pinchot Institute and, 73, 76; proposal by, 15–16; renovation and, 111, 112; resource analysts from, 144; venture with, 17
Conservation Hall of Fame Trail, 116
Conservation Handbook, A (Ordway), 42
Conservation Institute, 64, 65, 72
Conservation Legacy, 145
conservationism, 38, 48, 174, 176; practical, 133, 171–72
conservationists, 14, 90, 92, 154, 171, 174, 175; agenda of, 38
Consolidated Appropriations Act (2005), 152
Conte, Silvio, 82, 83, 116
Cortner, Hannah J., 145, 149; quote of, 137
Council on Environmental Quality, 144
Creative Act (1891), 155
Creative Services (National Wildlife Federation), 92
Crevecoeur, Guilaume-Michel Saint-John de, 26
Cristobal Colon, 161, 162, 163, 170, 204n16
Crow, Tom, 147

Dana, Samuel T., 66, 90, *104*
Daniels, Orville, 145
Darling, J. Ding, 90

Howard Heinz Foundation, 142
Hunt, Richard Morris, 12, 29, 127
Hyde, Tom, 164
hydropower, 57, 58

Ideal Farm, 169
inadvertent habitat pockets, 79
Industrial Revolution, 36
Information and Education, 115
"Invite Wildlife to Your Backyard," 91
Izaak Walton League v. Butz (1973), 146

Jatun Sacha, 161, 204n16
Jefferson, Thomas, 19
Job Corps, 80, 81
Johnson, Eastman, 127, 131
Johnson, Lady Bird, 69, 70, 189n11
Johnson, Leroy, 128, 130
Johnson, Lyndon B., 86, 134, 142,
   189n11; environmental protection
   and, 55; Great Society and, 60
Jorritsma, Jan, 170
Jorritsma family, 1–2, 169
Journal of Forestry, 48, 201n16

Kaplan, Rachel, 88
Kaplan, Stephen, 88
Kastner, Joel, *107*
Kastner, Mariana, *107*
Keeper of the National Register of
   Historic Places, 119
Kelley Family Foundation, 165
Kennedy, John F., 44, 86, 89, *102,* 142,
   186n13; assassination of, 50, 67, 73;
   conservation and, 54–60, 186n11;
   dedication and, 4, 20, 23, 25, 52, 134,
   155, 171, 174–75; described, 53, 185n3;
   environmentalism and, 55, 59; Forest
   Service and, 45–46; Grey Towers
   and, 2, 4, 8, 30; natural resources
   and, 46; nicknames for, 54; scientific
   management and, 58
Kennedy, Robert F.: assassination of, 80
Kids in the Woods, described, 193n23
Kimbell, Gail, 125–26, 193n23
King, Martin Luther, Jr., 80
Klein, William, 132, 199n40
Kotter, John, 130

Krawitz, Sidney, 140
Kuhn, Thomas, 175

Labounty, Paul, *108*
Lafayette, 29; chateau of, 13
Lake, Robert M., 115, 116
Land and Water Conservation Fund, 50
land management, 47, 48, 49, 50, 111, 113
*Land Stewardship in the Next Era of
   Conservation* (Sample), 156
Lands Nobody Wanted, The (Shands and
   Healy), 144
landscapes, 135, 174; managing, 79–80;
   regeneration of, 49
Larson, Geraldine, 125
Lassen Volcanic National Park, 57
Lassie—Forest Ranger, 89
Leading Change (Kotter), 130
Lehde, Norman, 19, 20
LeMaster, Dennis, 146
Leopold, Aldo, 37, 40, 46, 154, 157, 171
Letter Box, 128, 131
Lewis, James G., 125, 148
Library of Congress, Pinchot papers
   at, 118
Limited Test Ban Treaty, 59
Louv, Richard, 193n23
Love Canal, 124

MacArthur, Diana, 69
*Man and Nature: Earth as Modified by
   Human Action* (Marsh), 3, 154
*Man and Nature, or, Physical Geography as
   Modified by Human Action* (Marsh), 36
management, 6, 13, 80, 131, 150,
   156, 159; forest, 123, 137, 168, 172,
   205n26; land, 47, 48, 49, 50, 111, 113;
   landscape, 158; sustainable, 204n15;
   team-based, 130; top-down, 143
marinescapes, 174
Marris, Emma, 174
Marsh, George Perkins, 3, 36, 37, 40,
   154, 187n16
Mary Pinchot (schooner), 31–32
Massachusetts Audubon Society, 71
Masses (journal), 52
Master Facilities Site Plan, 133
Mater, Catherine: quote of, 154

McAlpin, David Hunter, 16
McArdle, Richard E., 15, 121–22
McDade, Joseph M., 113, 116, 117, 118, 122, 139, 152; charges against, 195n11; Grey Towers and, 134, 138, 140
McGuire, John R., 114, 118, 120
McKinley, William, 47
Mead, Margaret, 88
Mebus, Carole, 93–94
Medvetz, Thomas, 177n6
Melcher, John, 145
Metropolitan Tree Improvement Alliance (METRIA), 87
Meyer, Cort, 52
Meyer, Mary Pinchot, 18, 23, 52, 185n1
Milford, 3, 13, 15, 19, 21, 22, 26, 28, 29, 30, 36, 185n1; commercial prospects of, 27; described, 11; gifts to, 179n18; Kennedy in, 20, 44, 52, 64; Pinchot arrival in, 25
Milford and Oswego turnpike, 27
Milford Experimental Forest, 135, 160, 161
Miscellaneous Receipts Act, 152
Misty Fiords National Monument, 152
Moeller, George, 90
Monongahela National Forest, 49, 87
Monroe, James, 19
Mormon Tabernacle, Kennedy at, 57, 186n13
Morris Arboretum, 132, 199n40
Muir, John, 46, 129, 171
Multiple-Use Sustained-Yield Act (1960), 46
Murtaugh, William J., 119

Napoleon, 25, 26
National Conservation Association, 6, 33, 73
National Environmental Policy Act (1970), 50, 110, 175
National Forest Management Act (NFMA) (1976), 49, 50, 146, 147, 175
National Forest Resource Management, 50
National Forest System, 121, 152
national forests, 48; clear-cutting, 87; management of, 6; timber harvests in, 148
National Friends of Grey Towers, 132, 137, 138, 140, 144–45, 151; donations for, 139; fundraising by, 141
National Historic Landmarks, 6, 152, 178n8
National Historic Preservation Act (1966), 5, 118
National Historic Sites, 6, 152
National Historic Sites Act (1935), 118
National Institutes of Health, 82
National Leadership Team, 157
National Park Service, 2, 32, 50, 51, 59, 117; Grey Towers and, 134–35, 152
National Resources Council of America, 16, 41
National Resources Defense Council (NRDC), 33
National Review, 53
National Training Center, 114, 115
National Wildlife Federation, 90, 92, 94
National Wildlife, 90, 91, 92, 93
National Youth Conference for Natural Beauty and Conservation, 69
Natural Resource Conservation Service, 2
natural resources, 36, 46, 151, 155; conservation and, 52, 64; consumption of, 39, 68; focus on, 57; productivity of, 142
Natural Resources Council of America, 67
nature, 36, 176; balance of, 174; quality of life and, 193n23
Nature, 39
Nature Conservancy, 2, 169
Nelson, Gaylord, 56, 57, 59, 187n15
Nelson, Thomas C., 114, 115, 116
Neville, Leo (Bob), 123, 127, 129
New Deal, 38, 53, 86
New England River Basins Commission, 142
New Jersey Department of Environmental Protection, 169
New Perspectives, 147, 148, 149, 150
New York Foundation, 66
New York Times, 129

73; on national forest management, 6; notes to, 18; outbuildings by, 128; papers of, 118; politics and, 31, 52; public service and, 119; quote of, 11; Society of American Forests and, 157; timber famine and, 171; U.S. Senate and, 31
Pinchot, Gifford (Long Giff; Gifford Pinchot's nephew), 18, 178n16
Pinchot, Gifford Bryce, 8, 19, 23, 25, 30, 41, 66, 71, 73, *101*, *102*, 104, *106*, *107*, 112, 123; book by, 181n15; clear-cutting and, 49; Conservation Foundation and, 17, 76; death of, 135; dedication and, 24; disappointment for, 117; donation by, 2; education and, 64; environmental issues and, 32, 33; Grey Towers and, 3, 12, 65, 110, 121, 124; harvesting techniques and, 146; inheritance of, 13; Kennedy and, 53, 185n3; NRDC and, 33; Pacific voyage and, 31–32; public service of, 31–32; repairs by, 14; resource management and, 42; social change and, 158; Tocks Island and, 32–33
Pinchot, Gifford, III, 8, 28, *107*, 130, 135
Pinchot, James, 3, 11, 27, 29, 30, *96*, 154; art collection of, 65, 127; environmental convictions of, 127; financial success for, 28; Grey Towers and, 110, 139, 159; politics and, 3; Yale School of Forestry and, 135
Pinchot, Jeremy, *107*
Pinchot, Leila, *107*, 160
Pinchot, Libba, 130
Pinchot, Maria, 26, 27
Pinchot, Marianna, *107*
Pinchot, Mary Eno, 3, 11, 29, 30, *96*, 127, 165; marriage of, 28
Pinchot, Nancy (Peter's wife), 161
Pinchot, Nancy Pittman, 32, 129, 135, 197n32
Pinchot, Peter, 8, 29, 32, 48, *107*, *108*, 129, 135, 138, 157, 158, 160, 161, 163; on community-based forestry, 204n15; dedication and, 23; EcoMadera and, 162; on Grey Towers, 13; management

strategies of, 159; rededication and, 136
Pinchot, Rosamond, 198n32
Pinchot, Ruth Pickering, 12, 14, 17, 18, 52, 64, 65, *96*, 111; dedication and, 23; disappointment for, 117; Kennedy and, 21, 23, 52; Pinchot Institute and, 76; politics and, 53; visitors and, 123
Pinchot, Sarah (Sally) H., 13, 23, *106*, *107*, 182n15; on Gifford Bryce Pinchot, 31; Grey Towers and, 135; Kennedy and, 53
Pinchot (show), Hines and, 141
Pinchot Consortium, 90, 91, 116, 121, 124, 133; conference by, 193n23; conservation education and, 93, 137; Forest Service and, 5, 88, 141; funding for, 93; METRIA and, 87; symposia by, 85–86, 88
Pinchot Council, creation of, 145
Pinchot Estate, 113, 117, 133, 135
Pinchot Foundation, 124
Pinchot Institute for Conservation, 1, 2, 18–19, 145, 153; Forest Service and, 4; funding for, 5; survey by, 148
Pinchot Institute for Conservation Studies, 2, 8, 19, 46, 51, 60, 65, 69, 70, 85, 89, 90, *103*, 113, 117,124, 142, 145, 146, 147, 151, 160, 166, 167; activism of, 170; anniversary celebration for, 135; appeal of, 177n6; approach of, 84; collaboration by, 161; Common Waters and, 168; conservation and, 43, 75, 171–72; Conservation Foundation and, 73, 76; debates with, 176; dedication of, 21, 24, 25, 35–36, 55, 171, 174–75; EcoMadera and, 163; economic potential of, 19; ecosystem management and, 151; education and, 74; environment and, 75; establishment of, 36, 80; forest management and, 150; Forest Service and, 4, 73, 74, 76, 137, 138, 151; funding for, 92, 93; Grey Towers and, 8, 25, 77, 137, 141, 142; inaugural board of, *104*; insight/foresight of, 64; Kennedy at, 174–75;

leadership of, 7; Memorandum of Understanding with, 165; Milford Experimental Forest and, 159; mission of, 121, 128, 167; New Perspectives and, 149; principles of, 8, 156; purpose of, 74–75; reinvention of, 6, 80; resource management and, 42; success for, 63; water pollution and, 86; watersheds and, 157

Pinchot Institute for Environmental Forestry, 112–13

Pinchot Institute of Environmental Forestry Research (PIEFR), 80, 82, 83, 85

Pinchot Institute System for Environmental Forestry Studies, The (report), 84

Pleistocene, 173

"Plunder or Plenty" (Ordway), 39–40

Pocono Mountains, 159, 168

Pocono Record, 120

Point Reyes, 59

politics, 3, 31, 143, 177n6

pollution, 36; air, 81; vulnerability to, 81; water, 51, 86, 143

population: food supply and, 183n16; growth, 172

Population Bomb, The (Ehrlich and Ehrlich), 142

Port Jervis, 27–28

Port Jervis Union-Gazette, quote from, 21

preservation: ecosystem, 174; historic, 119, 123; wilderness, 175

Primer of Forestry (Pinchot), 161

Progressive Era, 6, 38, 50, 86

Prosperity beyond Tomorrow (Ordway), 42

Przedvorski, Julia, 163

Public Works Administration, 86

Pyles, Hamilton, 15

Quiet Crisis, The (Udall), 187n16

"Raccoons on the Roof" (Thomas and DeGraaf), quote from, 78

rainforests, 39, 155

Rains, Michael, 130, 153

Reagan, Ronald, 125, 141, 146; budget cuts by, 5, 93, 123; natural resources and, 129

Recycled Treated Wastewater and Sludge through Forest and Cropland, 87

reforestation, 81, 82, 204n16

refuge-in-miniature, 91, 92

Regence Blue Cross Blue Shield, 165

Regional Training Center, 114

Reilly, William K., 144, 148

Resler, April, 116

Resler, Rexford, 114–15, 116, 117

resource management, 42, 149, 155, 172, 174; dismantling, 150; promoting, 43

Resources and the American Dream (Ordway), 40

Resources for the Future, 41, 66, 150

restoration, 91, 111, 112, 169–70

Rio Canandé, 161, 163, 204n16

Rio Verde Canandé, 162

Road to Survival, The (Vogt), 38, 39

Robertson, Dale, 144, 147, 148, 149

Robertson, Thomas, 38, 201n16

Robinson, Dale, 22

Rockefeller, Laurence, 16, 37, 66, *104*; Osborn/Ordway and, 41

Rogers, Amy, 204n16

Roget, Einar L., 115

Roosevelt, Franklin D., 90; conservation and, 52, 54, 56, 58; foreign policy of, 53

Roosevelt, Theodore, 37, 47, *98*, 184n7; conservation and, 48, 52, 54, 56, 58, 59; national forests and, 48; Pinchot and, 31, 46; protection/support of, 45

Roth, Charles E., 71

Russell Cave, 55

SAF. *See* Society of American Foresters

Sajdak, Nathaniel, 2

Salvation Army, 14

Sample, V. Alaric, 7, *108*, 144, 146, 150, 156, 206n1; Brannon and, 151; Pinchot Institute and, 145

Sampson, R. Neil, 143

San Juan-Navaho Indian Project, 56

Sand County Almanac, A (Leopold), 154

Saturday Review, Ordway in, 39–40

Sawkill Creek, 3, 12, 32, 159